NELSON VICscience

Claire Wallace

Emma Bleazby

Gabrielle Painter

SKILLS WORKBOOK

psychology

VCE UNITS ① + ②

4E

Nelson

VICscience Psychology VCE Units 1 & 2 Skills Workbook
4th Edition
Claire Wallace
Emma Bleazby
Gabrielle Painter
ISBN 9780170465045

Publisher: Eleanor Gregory
Content developer: Katherine Roan
Project editors: Felicity Clissold, Alex Chambers, Alana Faigen
Copyeditor: Holly Proctor
Proofreader: Nick Tapp
Series text design: Ruth Comey (Flint Design)
Series cover design: Emilie Pfitzner, Everyday Ambitions
Series designer: Cengage Creative Studio
Permissions researcher: Liz McShane
Production controllers: Karen Young, Alex Chambers
Typeset by: Lumina Datamatics
Any URLs contained in this publication were checked for currency during the production process. Note, however, that the publisher cannot vouch for the ongoing currency of URLs.

Acknowledgements
Extracts from the VCE Psychology Study Design (2023-2027) are used by permission, © VCAA. VCE® is a registered trademark of the VCAA. The VCAA does not endorse or make any warranties regarding this study resource. Current VCE Study Designs, past VCE exams and related content can be accessed directly at www.vcaa.vic.edu.au.

We would like to acknowledge the following for permission to reproduce copyright material - images for each chapter opener:
Ch 1: iStock.com/SolStock
Ch 2: iStock.com/StefaNikolic
Ch 3: Alamy Stock Photo/Shiiko Alexander
Ch 4: Adobe Stock/sutadimages
Ch 5: Alamy Stock Photo/Cultura Creative RF
Ch 6: iStock.com/Halfpoint
Ch 7: Adobe Stock/oneinchpunch
Ch 8: iStock.com/SDI Productions
Ch 9: iStock.com/Imgorthand
Ch 10: Shutterstock.com/MrslePew
Ch 11: Adobe Stock/Rachael Arnott

© 2023 Cengage Learning Australia Pty Limited

For product information and technology assistance,
in Australia call **1300 790 853**;
in New Zealand call **0800 449 725**

For permission to use material from this text or product, please email
aust.permissions@cengage.com

ISBN 978 0 17 046504 5

Cengage Learning Australia
Level 5, 80 Dorcas Street
Southbank VIC 3006 Australia

Cengage Learning New Zealand
Unit 4B Rosedale Office Park
331 Rosedale Road, Albany, North Shore 0632, NZ

For learning solutions, visit **cengage.com.au**

Printed in China by 1010 Printing International Limited.
1 2 3 4 5 6 7 26 25 24 23 22

Contents

Introduction

Psychology is a multifaceted discipline that seeks to describe, explain, understand and predict human behaviour and mental processes (VCAA VCE *Psychology Study Design 2023–2027*, page 6). Like the study of any other science there is key knowledge and terminology that you need to know and understand and be able to use appropriately. However, no study of Psychology would be complete without also addressing the key science skills. The study of Psychology is not just about learning content, it is also about developing, using and demonstrating the skills that enable you to fully understand, experience and engage with the subject. It is about learning to think and work like a scientist.

Seven key science skills have been mandated by the Victorian Curriculum Assessment Authority (VCAA) across all VCE science subjects. These skills are transferable across subjects as well as being examinable in the VCE exam. Developing these key science skills means that you will be able to:

- develop aims and questions, formulate hypotheses and make predictions
- plan and conduct investigations
- comply with safety and ethical guidelines
- generate, collate and record data
- analyse and evaluate data and investigation methods
- construct evidence-based arguments and draw conclusions
- analyse, evaluate and communicate scientific ideas.

(VCAA VCE *Psychology Study Design 2023–2027*, pages 11–12)

Each of the key science skills listed above is broken up into multiple sub-skills. The mapping provided on pages vii–x of this workbook allows you to see how these skills and sub-skills have been addressed in this workbook.

This workbook is full of activities that have been carefully crafted to enable you to consolidate your knowledge on a topic and to develop, use and demonstrate key science skills. Developing any skill takes time and practice; the key science skills in this book have been introduced in a graduated way starting with **practising** skills that you will have met in previous years of science study. As you gain proficiency and confidence, you will go on to **reinforce** newer and more complex skills. Then there are the new skills requiring an increased level of proficiency and thanking that you will **develop** during the course.

 Shows you activities that require previously introduced skills and will require practise as you work through the activities.

 Shows activities that will build on previously introduced skills.

 Shows activities that introduce a new skill or skills that require development and challenge you at a high level of proficiency.

This workbook can be used with any VCE Psychology resource that covers the VCAA VCE *Psychology Study Design 2023–2027*. It has been mapped to VICscience Psychology Units 1 and 2 using icons in both the workbook and the student textbook. Icons have been placed in the student textbook to indicate the best place to undertake each activity.

The major headings in the workbook match the major headings in VICscience Psychology Units 1 and 2. Applicable key knowledge is listed under each of the major headings. Skill activities have the applicable key science skills listed that students will be using or demonstrating.

Enjoy your study of VCE Psychology and take the time to develop, use and demonstrate the key science skills that are an integral part of this course.

Study design grid

Key science skill	VCE Psychology Units 1–2	1	2	3	4	5	6	7	8	9	10	11
Develop aims and questions, formulate hypotheses and make predictions	identify, research and construct aims and questions for investigation	1.1.1							8.1.4 8.2.2 8.3.1 8.4.1	9.2.2	10.2.1 10.2.2	11.1.1
	identify independent, dependent, and controlled variables in controlled experiments	1.1.1 1.4.5	2.1.1 2.3.1 2.4.2		4.4.6	5.3.3			8.1.4 8.2.2		10.2.2	11.1.2
	formulate hypotheses to focus investigation	1.1.1 1.4.5			4.4.6	5.3.3			8.1.4 8.2.2 8.3.1 8.4.1		10.2.2	11.1.2
	predict possible outcomes of investigations											
Plan and conduct investigations	determine appropriate investigation methodology: case study; classification and identification; controlled experiment (within subjects, between subjects, mixed design); correlational study; fieldwork; literature review; modelling; product, process or system development; simulation	1.2.1 1.3.2					6.1.2					11.1.1 11.1.3
	design and conduct investigations; select and use methods appropriate to the investigation, including consideration of sampling technique (random and stratified) and size to achieve representativeness, and consideration of equipment and procedures, taking into account potential sources of error and uncertainty; determine the type and amount of qualitative and/or quantitative data to be generated or collated	1.1.2 1.1.4 1.1.5 1.3.1	2.1.3			5.3.2						11.1.4 11.1.5
	work independently and collaboratively as appropriate and within identified research constraints, adapting or extending processes as required and recording such modifications											

Category	Skill	1	2	3	4	5	6	7	8	9	10	11
Comply with safety and ethical guidelines	demonstrate ethical conduct and apply ethical guidelines when undertaking and reporting investigations		2.4.2		4.4.6		6.2.1		8.1.4 8.2.2 8.3.1 8.4.1	9.2.2		
	demonstrate safe laboratory practices when planning and conducting investigations by using risk assessments that are informed by safety data sheets (SDS), and accounting for risks										10.2.3	
	apply relevant occupational health and safety guidelines while undertaking practical investigations											
Generate, collate and record data	systematically generate and record primary data, and collate secondary data, appropriate to the investigation	1.1.3 1.4.1		3.1.3				7.1.1	8.3.1	9.2.1	10.2.1 10.2.2 10.2.3	
	record and summarise both qualitative and quantitative data, including use of a logbook as an authentication of generated or collated data											
	organise and present data in useful and meaningful ways, including tables, bar charts and line graphs	1.4.1				5.4.3					10.2.1	11.1.6
Analyse and evaluate data and investigation methods	process quantitative data using appropriate mathematical relationships and units, including calculations of percentages, percentage change and measures of central tendencies (mean, median, mode), and demonstrate an understanding of standard deviation as a measure of variability	1.4.1 1.4.4	2.3.3	3.1.3		5.1.2		7.1.1	8.3.1 8.4.1		10.2.1	
	identify and analyse experimental data qualitatively, applying where appropriate concepts of: accuracy, precision, repeatability, reproducibility and validity; errors; and certainty in data, including effects of sample size on the quality of data obtained	1.4.2 1.4.5				5.1.2 5.4.3			8.2.3		10.2.1	11.1.7
	identify outliers and contradictory or incomplete data	1.4.4										
	repeat experiments to ensure findings are robust											
	evaluate investigation methods and possible sources of error or uncertainty, and suggest improvements to increase validity and to reduce uncertainty	1.4.3					6.1.3					11.2.1

9780170465045

		1	2	3	4	5	6	7	8	9	10	11
Construct evidence-based arguments and draw conclusions	distinguish between opinion, anecdote and evidence, and scientific and non-scientific ideas					5.4.4						
	evaluate data to determine the degree to which the evidence supports the aim of the investigation, and make recommendations, as appropriate, for modifying or extending the investigation											11.2.1
	evaluate data to determine the degree to which the evidence supports or refutes the initial prediction or hypothesis				4.4.6				8.3.1 8.4.1			11.2.1
	use reasoning to construct scientific arguments, and to draw and justify conclusions consistent with evidence base and relevant to the question under investigation						6.1.2 6.1.3 6.2.1		8.3.2		10.3.2	11.2.1
	identify, describe and explain the limitations of conclusions, including identification of further evidence required	1.4.2 1.4.5					6.1.3		8.1.4 8.2.2			11.2.1
	discuss the implications of research findings and proposals, including appropriateness and application of data to different cultural groups and cultural biases in data and conclusions		2.1.1	3.2.1					8.2.3 8.4.1	9.2.2		11.2.1

9780170465045

Analyse, evaluate and communicate scientific ideas	use appropriate psychological terminology, representations and conventions, including standard abbreviations, graphing conventions and units of measurement	1.1.6		3.1.3 3.2.1		5.4.2					
	discuss relevant psychological information, ideas, concepts, theories and models and the connections between them	1.1.3 1.1.4		3.1.1 3.1.2 3.1.4 3.2.2 3.3.1 3.3.2 3.4.1 3.4.2 3.4.3 3.4.4	4.1.1 4.1.2 4.2.1 4.2.2 4.2.3 4.2.4 4.2.5 4.2.6 4.3.1 4.3.2 4.3.3 4.3.4 4.4.1 4.4.2 4.4.3 4.4.4 4.4.5	5.1.1 5.1.2 5.1.3 5.2.1 5.2.2 5.2.3 5.2.4 5.2.5 5.2.6 5.3.1 5.4.1 5.4.2	6.1.2 6.1.3	7.1.2 7.1.3 7.1.4 7.1.5 7.2.1 7.2.2 7.3.1 7.4.1 7.4.2 7.4.3 7.4.4 7.4.5	8.1.1 8.1.2 8.1.3 8.1.4 8.2.1 8.2.2 8.2.3 8.2.4 8.2.5 8.2.6 8.3.2	9.1.1 9.1.2 9.1.3 9.3.1 9.3.2 9.4.1 9.4.2	10.1.1 10.1.2 10.3.1 10.3.3 10.3.4
	analyse and explain how models and theories are used to organise and understand observed phenomena and concepts related to psychology, identifying limitations of selected models/ theories		2.1.2 2.2.1 2.2.2 2.3.1 2.3.2 2.3.4 2.3.5 2.4.1	3.1.1 3.1.3 3.2.1 3.4.2				7.1.3 7.1.4 7.1.5 7.2.1	8.2.3 8.3.2	9.1.1 9.1.2 9.1.3 9.2.1 9.3.1 9.3.2 9.4.1 9.4.2	
	critically evaluate and interpret a range of scientific and media texts (including journal articles, mass media communications, opinions, policy documents and reports in the public domain), processes, claims and conclusions related to psychology by considering the quality of available evidence				4.4.4	5.1.3 5.4.4	6.1.2 6.1.3 6.2.1	7.2.1 7.4.4		9.2.2	
	analyse and evaluate psychological issues using relevant ethical concepts and principles, including the influence of social, economic, legal and political factors relevant to the selected issue	1.4.6 1.4.7	2.1.4	3.3.2 3.4.3			6.2.1				
	use clear, coherent, and concise expression to communicate to specific audiences and for specific purposes in appropriate scientific genres, including scientific reports and posters	1.1.6					6.1.1				
	acknowledge sources of information and assistance, and use standard scientific referencing conventions	1.1.6		3.4.4		5.1.3					

9780170465045

Scientific research methods

1.1 The process of psychological research investigations

1.1.1 Formulating hypotheses

> **Key science skills**
> Develop aims and questions, formulate hypotheses and make predictions
> * identify, research and construct aims and questions for investigation
> * identify independent, dependent, and controlled variables in controlled experiments
> * formulate hypotheses to focus investigation
>
>
> Develop

PART A

For each of the scenarios provided complete the following:

1 identify the independent variable (IV)
2 identify the dependent variable
3 formulate your own research hypothesis.

> Remember a hypothesis is a prediction about the relationship between the IV and DV.

SCENARIO 1

Dr Hunter wants to research the influence of caffeine on ability to memorise in middle-aged men. He gives his experimental group four cups of coffee a day and then measures the number of words they recall on a series of short-word recollection tests. The control group completes the same tests but does not consume any caffeine.

1 Independent variable:

_____

2 Dependent variable:

_____

3 Hypothesis:

_____

_____

SCENARIO 2

A teacher was disappointed with the mid-year exam results achieved by her class and wanted to improve her students' learning for the end-of-year exam. During the second semester, she used a variety of coloured markers on her whiteboard, instead of using only black, as she had done in the first semester.

1 Independent variable:

_____

2 Dependent variable:

_____

3 Hypothesis:

_____

_____

SCENARIO 3

The Ferrari team wanted to see if their drivers' reaction times would improve if they fitted their cars with gauges that gave digital displays, rather than the analogue dial and needle arrangement.

1 Independent variable:

_____

2 Dependent variable:

_____

3 Hypothesis:

_____

_____

SCENARIO 4

Dr Woodford, a psychologist, was concerned about levels of stress experienced by local council employees. He wanted to investigate whether he could lower their stress levels, as measured by a self-report inventory, by medicating them daily with a sedative. A control group completed the same inventory but did not take any of the medication.

1 Independent variable:

_____

2 Dependent variable:

_____

3 Hypothesis:

_____

_____

SCENARIO 5

For his PhD, a university student investigates the effects of sleep deprivation on adults' ability to solve simple mathematical problems. He keeps Group 1 awake for 12 hours, Group 2 awake for 24 hours, Group 3 awake for 36 hours and Group 4 awake for 48 hours. He then gives all participants a series of basic arithmetic problems to solve.

1 Independent variable:

_____

2 Dependent variable:

_____

3 Hypothesis:

_____

_____

SCENARIO 6

A group of researchers wanted to investigate heart rate and blood pressure in groups of children watching violent and non-violent films.

1 Independent variable:

_____

2 Dependent variable:

_____

3 Hypothesis:

_____

_____

SCENARIO 7

Professor Smith is examining the effect of hunger as motivation in rats trying to run through a maze. He uses two groups of rats, and times how long it takes members of each group to successfully complete the maze to find a piece of cheese. Professor Smith feeds the rats in one group before they enter the maze, but does not feed the rats in the other group until they have completed the maze.

1 Independent variable:

_____

2 Dependent variable:

_____

3 Hypothesis:

_____

_____

PART B

In the following experiment outlines, list the independent variable, dependent variable, how each is measured, experimental group, control group and hypothesis.

1 The aim of the experiment was to investigate the effect of study on exam performance. There were four groups of participants involved in the study. One group did not study at all, one group studied for 4 hours, one group studied for 10 hours and one group studied for 20 hours prior to the exam. The researcher administered the exam to each of the participants with results recorded.

    a Independent variable:

_____

    b How is the independent variable measured?

_____

_____

    c Dependent variable:

_____

    d How is the dependent variable measured?

_____

_____

    e Experimental group(s)

_____

_____

f Control group

_____

g Hypothesis

_____

_____

2 The aim of the experiment was to investigate the effect that passengers have on one's driving performance. Each group of drivers performed a driving test on a course that had obstacles on it. One group of participants had no passengers in the car, one group had one passenger in the car and another group had four passengers in the car while completing the test. The researcher recorded the number of driving errors that each participant made.

a Independent variable:

_____

b How is the independent variable measured?

_____

_____

c Dependent variable:

_____

d How is the dependent variable measured?

_____

_____

e Experimental group(s)

_____

_____

f Control group

_____

g Hypothesis

_____

_____

1.1.2 Sampling procedures

Key science skills

Plan and conduct investigations
- design and conduct investigations; select and use methods appropriate to the investigation, including consideration of sampling technique (random and stratified) and size to achieve representativeness, and consideration of equipment and procedures, taking into account potential sources of error and uncertainty; determine the type and amount of qualitative and/or quantitative data to be generated or collated

Develop

MATERIALS

- scissors
- glue

Extraneous variables are not great for our results.

INSTRUCTIONS

1 For each type of sampling in Table 1.1, find the following in Table 1.2:
   a the definition of the sampling technique
   b an advantage of the sampling technique
   c a disadvantage of the sampling technique
   d an example of a use that best suits the sampling technique.
2 Cut out the boxes in Table 1.2 and glue them into the appropriate cells in Table 1.1.

Table 1.1 The types of sampling

Type of sampling	Definition	Advantage	Disadvantage	Example of use
Convenience sampling				
Random sampling				
Stratified sampling				
Random-stratified sampling				

Table 1.2 Descriptions of the types of sampling

Members of the population are broken into groups, or strata, based on particular characteristics; a proportionate number of members in each group is then selected for the sample.	The sample may not be representative of the population.	It is time- and cost-effective to select a large sample.	Separate a year level into male and female students and then pull out 10 boys' names and 10 girls' names from two separate hats.
The sample is likely to be biased.	Sample the first 20 people who enter the library.	Sample students by writing down all of their names and then drawing five names from a hat.	Every member of the population has an equal chance of being selected for the sample.
Participants are selected for the sample based on the ease of access and selection.	The sample has a proportionate number of participants representing each characteristic in the population.	The sample is representative of the population.	Members of the population are broken into groups, or strata, based on particular characteristics; a proportionate number of members in each group is then randomly selected for the sample.
It takes a lot of resources (time and money) to select a sample.	Although the sample is a representative sample, the participants selected from each characteristic may be biased.	The sample is very easy to obtain.	Separate a year level into male and female students and then select the first females and males that you see.

Table 1.2 Descriptions of the types of sampling

1.1.3 Applying random sampling

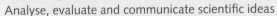

Key science skills

Generate, collate and record data
- systematically generate and record primary data, and collate secondary data, appropriate to the investigation

Analyse, evaluate and communicate scientific ideas
- discuss relevant psychological information, ideas, concepts, theories and models and the connections between them

Develop

MATERIALS

- a large container of coloured beads (enough for five groups of 50)

INSTRUCTIONS

1 Randomly divide the container so that each group has a sample of 50 beads.
2 Sort the 50 beads into colours.
3 Count how many of each colour and record the numbers in Table 1.3.
4 Convert the data into a percentage and record this in the table.
5 Combine the data from the whole class and get the actual totals and percentages of each colour.

Table 1.3 Group and class data

Colours									Total
Group									50
Group (%)									
Class									
Class (%)									

1 Comment on how representative the sample was to the whole population.

1.1.4 Sampling flowchart

Key science skills

Plan and conduct investigations
- design and conduct investigations; select and use methods appropriate to the investigation, including consideration of sampling technique (random and stratified) and size to achieve representativeness, and consideration of equipment and procedures, taking into account potential sources of error and uncertainty; determine the type and amount of qualitative and/or quantitative data to be generated or collated

Analyse, evaluate and communicate scientific ideas
- discuss relevant psychological information, ideas, concepts, theories and models and the connections between them

Develop

Provide definitions for each of the stages shown in the flowchart in Figure 1.1. Add examples of advantages and disadvantages where required. It may help to increase the size of this flowchart by photocopying it onto an A3 sheet of paper.

Figure 1.1 A psychological research flowchart

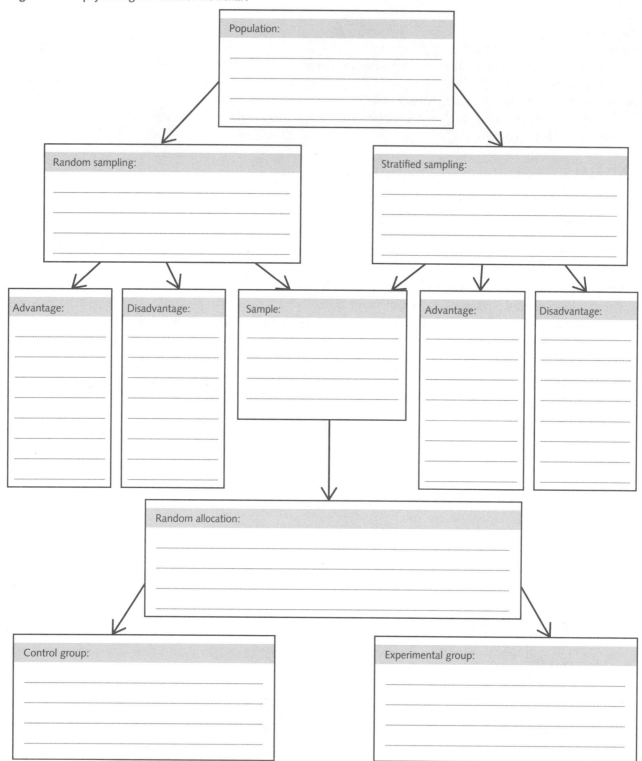

1.1.5 Ways to obtain data

Key science skills

Plan and conduct investigations

- design and conduct investigations; select and use methods appropriate to the investigation, including consideration of sampling technique (random and stratified) and size to achieve representativeness, and consideration of equipment and procedures, taking into account potential sources of error and uncertainty; determine the type and amount of qualitative and/or quantitative data to be generated or collated

Develop

Use the terms in Table 1.4 to fill in the blanks in the following paragraphs.

Table 1.4 Data terms

subjective	interview	observational	self-fulfilling	observer	objective
quantitative	imaging	qualitative	questionnaire	case study	compare
rating scale	open-ended	naturalistic	generalise	self-report	sample

To conduct an experiment in psychology, you need to obtain data from the _____ that is representative of the population of interest for the particular study. This can be done in many ways. A common method is to ask people to reveal information about themselves; this is a(n) _____ technique. This technique could involve participants completing a series of questions or responding to statements by means of a(n) _____. One very common form of this is a(n) _____, where a series of statements are produced and participants are required to circle whether they strongly agree, agree, neither agree nor disagree, disagree or strongly disagree with each statement.

This technique of data collection could also involve meeting one-on-one with a participant and asking them about various issues in a(n) _____. One advantage to this method of data collection is that questions can be _____, which allows participants free choice in how to answer a question. This means that the responses are descriptive and detailed, and that participants can explain how they feel about a chosen topic. This sort of data is known as _____ data.

An advantage of collecting this sort of data is that a lot of detail can be obtained; however, it can make it difficult to _____ the data with other information. Alternatively, collecting _____ data makes comparing and graphing data very easy. This type of data can be numerical or categorical. One disadvantage is that it is restrictive in terms of participants' responses.

When using participant reports as a means of obtaining data, many problems may be encountered. One problem is that self-reports rely on personal interpretation, and the participant can provide a response that they believe the experimenter wants. To eliminate the _____ prophecy, experimenters can watch individuals in a(n) _____ setting. This is known as a(n) _____ study. A limitation to using this experimental method is that the data is interpreted by an experimenter, who may report what they expect to see. This is _____ bias. It means that the data is _____ in nature, whereas data that can be directly measured is _____ data.

An in-depth investigation conducted on a single person or small group of people is a(n) _____. This type of study provides very detailed information, but it is difficult to _____ the findings to the population.

Modern technologies, such as brain _____ techniques, have allowed data to be collected in a more objective manner. Large amounts of data can be obtained via brain scans, enabling us to learn more about the brain and behaviour.

1.1.6 Features of a scientific research report

Key science skills
Analyse, evaluate and communicate scientific ideas
- use appropriate psychological terminology, representations and conventions, including standard abbreviations, graphing conventions and units of measurement
- use clear, coherent, and concise expression to communicate to specific audiences and for specific purposes in appropriate scientific genres, including scientific reports and posters
- acknowledge sources of information and assistance, and use standard scientific referencing conventions

Develop

PART A

MATERIALS

- scissors
- glue

INSTRUCTIONS

Table 1.5 contains the names of the different sections of a scientific report, along with their descriptions, which have been jumbled. Cut out the sections and descriptions, and place them, correctly in order, in Table 1.6.

Table 1.5 Jumbled sections of a scientific research report and descriptions

Abstract	This is a subsection in the Method that lists the necessary resources to conduct the experiment.
Method	This is one sentence at the very beginning of a scientific research report that describes the potential cause-and-effect relationship that will be investigated.
Procedure	This section of a scientific report is broken into three subsections that describe how the experiment was conducted.
Title	This section of a scientific report is a collection of any materials and/or raw data used in the scientific report, featured in their entire form.
Participants	This is a subsection in the Method that provides all of the steps involved in replicating the experiment.
Introduction	This section of a scientific report is a brief summary of all aspects of the investigation.
References	This section of a scientific report displays a visual and written representation of the analysed data obtained in the experiment.
Materials	This is a subsection in the Method that lists the details of the people used in the research.
Discussion	This section of a scientific report features background information on the topic being investigated as well as past research, the aim, hypothesis, IV and DV.
Results	This section of a scientific report includes information supporting or refuting the hypothesis, discussing extraneous variables and drawing conclusions.
Appendix	This section of a scientific report lists all of the resources that were consulted when writing the report.

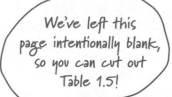

Table 1.6 Sections of a scientific research report and descriptions

Section	Description

Table 1.6 Sections of a scientific research report and descriptions

PART B

Indicate whether each statement is true or false by placing a tick in the correct column.

Statement	True	False
1 A scientific research report should be written in past tense.		
2 The results should be discussed in full detail in the results section.		
3 The title of the book appears first when referencing using APA format.		
4 The aim appears before the hypothesis in a scientific report.		
5 When graphing, both axes should be fully labelled.		
6 A research report should be scientific in style, not personal.		
7 A conclusion should always be made when conducting research.		
8 The participants should be mentioned in the materials section.		
9 When referencing quotes in the body of the text, only the author's surname and year of publication are required.		
10 When drawing a graph, no title is necessary because the section is already titled 'Results'.		

Scientific investigation methodologies

1.2.1 Research methodologies and investigation aims

Key science skills
Plan and conduct investigations
- determine appropriate investigation methodology: case study; classification and identification; controlled experiment (within subjects, between subjects, mixed design); correlational study; fieldwork; literature review; modelling; product, process or system development; simulation

Develop

Read the investigation aims stated in Table 1.8 and determine which research methodology from Table 1.7 would be best used.

Table 1.7 Examples of research methodologies

Case study	Product, process or system development	Longitudinal study
Correlational study	Fieldwork	Cross-sectional study
Classification and identification	Controlled experiment	

Table 1.8 Choosing appropriate research methodologies

Methodology	Investigation aim
	To analyse secondary school students' distress levels with regard to their year level.
	To investigate the cognitive abilities of Mr Kim Peek, who was born with a malformed cerebellum and without a corpus callosum.
	To determine if there are different sources of motivation to succeed in adolescents.
	To create sensory objects that assist children with autism to regulate their behaviour.
	To determine if there is a relationship between hours spent on social media and perceived stress.
	To determine factors that give Aboriginal and Torres Strait Islander children the best start in life to grow up to measure high on wellbeing criteria.
	To study children at play by observing their behaviour in a playgroup in Melbourne.
	To determine the effect of a new class of antidepressant on the amount of REM sleep experienced during each sleep cycle.

1.3 The controlled experiment in detail

1.3.1 Extraneous variables

Key science skills
Plan and conduct investigations
- design and conduct investigations; select and use methods appropriate to the investigation, including consideration of sampling technique (random and stratified) and size to achieve representativeness, and consideration of equipment and procedures, taking into account potential sources of error and uncertainty; determine the type and amount of qualitative and/or quantitative data to be generated or collated

Develop

For each scenario complete the following:

1 identify a possible extraneous variable that may affect the research

2 suggest a way to prevent the extraneous variable occurring that you could apply to the scenario so that it does not end up being a confounding variable.

SCENARIO 1

Maddi was conducting research to investigate the effect of music on study. She gave her participants a list of names starting with 'A' and asked them to learn it in silence. After 1 minute, they wrote down all the names they could recall. She then handed the same group of participants another list of names starting with 'A', and she played music as they were learning the names. After 1 minute, they wrote down all the names they could recall.

Extraneous variables are not great for our results.

1 Extraneous variable:

2 A way to prevent the extraneous variable:

SCENARIO 2

A teacher decides to give half the students in his class a new 'superfood' he has discovered, which is intended to improve their intelligence.

The remaining students in the class receive no superfood and are told they will have to compete academically as best they can.

1 Extraneous variable:

2 A way to prevent the extraneous variable:

SCENARIO 3

Psychville Basketball Club is trying to test the effectiveness of a new training program on shooting accuracy. They put their 10 tallest players through the training program and then asked them to shoot 100 balls from the three-point line. Their 10 shortest players did not complete the training program but also shot 100 balls from the three-point line. The shooting accuracy of the groups was then compared.

1 Extraneous variable:

2 A way to prevent the extraneous variable:

SCENARIO 4

Anh is a university graduate researching whether caffeine can improve driving ability. She is hoping that this is the case so that she can write a thesis on it. Group A ingests caffeine before taking a simulated driving test, and Group B does not ingest any caffeine before taking the same simulated driving test. As members of Group A are conducting their driving test, Anh actively encourages them. When members of Group B conduct their driving test, Anh stands over them and points out their errors.

1 Extraneous variable:

_____

2 A way to prevent the extraneous variable:

_____

_____

_____

1.3.2 Research designs

Key science skills

Plan and conduct investigations

- determine appropriate investigation methodology: case study; classification and identification; controlled experiment (within subjects, between subjects, mixed design); correlational study; fieldwork; literature review; modelling; product, process or system development; simulation

Develop

PART A

Define the research designs in Table 1.9. Outline the strengths and weaknesses of each.

Table 1.9 Research designs

Design	Definition	Strengths	Weaknesses
Within groups			
Between groups			
Mixed design			

When conducting an investigation, there are many different research designs to choose from.

PART B

Determine the research design being used in each scenario below.

1 A convenience sample of 20 Year 12 students was tested for concentration 30 minutes prior to eating their lunch and then 30 minutes after they had eaten their lunch. The test required them to read a short passage of text and identify 10 simple grammatical errors as quickly as possible. The time taken to identify all errors was measured in seconds._____

2 To determine the effect of texting on driving skill, orange traffic cones were set up in a circuit in a car park. A group of students were tested on the circuit, once while receiving and sending texts, and one without texting. Skill was measured by the number of cones hit while driving each circuit. _____

3 To determine whether childhood participation in sports is related to better self-esteem in adolescents, a representative sample of 100 adolescents with a history of sports participation were given a questionnaire on self-esteem and their scores were averaged. Another representative sample of 100 adolescents were also given this questionnaire, but they were selected with no history of sports participation. _____

4 Orlando wanted to investigate the effect of sleep and caffeine consumption on test performance. He allocated his sample into two groups, students who drank two cups of coffee and those who did not drink any coffee. For each of these conditions he further allocated his sample by number of hours slept; 3 hours, 6 hours and 9 hours. An example of his analysis includes the results of those who drink coffee and those that don't as well as those that drink coffee and the number of hours slept. _____

5 A psychology teacher wanted to test whether creating a mind map of a topic prior to testing would improve test results. He randomly allocated his students into two groups. One group was required to complete a mind map using a specific process in addition to their usual study techniques. The other group studied as usual. Test results were compared. _____

6 A researcher is studying the effect of a particular drug on nightmares in veterans with post-traumatic stress disorder (PTSD). A sample of veterans who were diagnosed with PTSD were asked to keep count of their nightmares for one month. They were then given the medication and again asked to record the number of nightmares for a month. They gave informed consent to receiving either the drug or a placebo. _____

7 Participants were divided into two groups. One group flew a flying simulator after consuming one standard drink; the other group did not consume any alcohol. Further to that, half the participants in each group were experienced pilots with over 10 years' experience, the other half had only just received their licence. Reaction time was measured. _____

8 Raj wants to determine whether smiling people are perceived as more intelligent than people who don't smile.
_____

9 Sarah wants to do an experiment looking at whether there are gender differences with sleep deprivation and on driving ability. _____

10 Researchers want to test the effects of a memory-enhancing drug. Participants are given one of three different drug doses, and then asked to either complete a simple or complex memory task. The researchers note that the effects of the memory drug are more pronounced with the simple memory tasks, but not as apparent when it comes to the complex tasks._____

11 Below is a table of results for an experiment.
   a What type of design was used to obtain these results?
   _____
   _____

Table 1.10 Effect of mood and type of word on recall

		Pleasant words	Unpleasant words
Mean recall			
Groups	Happy mood	70	23
	Sad mood	48	35

b Draw a graph of these results below.

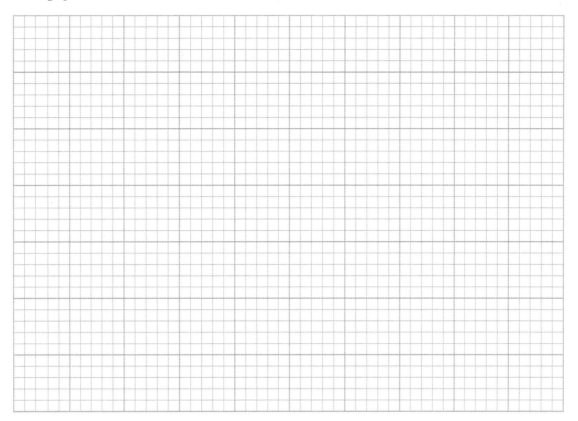

1.4   Analysing and evaluating research

1.4.1 Collecting and collating data

Key science skills

Generate, collate and record data

- systematically generate and record primary data, and collate secondary data, appropriate to the investigation
- organise and present data in useful and meaningful ways, including tables, bar charts and line graphs

Analyse and evaluate data and investigation methods

- process quantitative data using appropriate mathematical relationships and units, including calculations of percentages, percentage change and measures of central tendencies (mean, median, mode), and demonstrate an understanding of standard deviation as a measure of variability

Develop

PART A

MATERIALS

- tape measure

INSTRUCTIONS

1 Follow the steps to measure your height and complete the tasks and questions that follow.
2 Measure your height in centimetres (without shoes) and write it on the board in your classroom or on a blank piece of paper if you are working outside of your classroom. All members of the class should do this.
3 Record all the heights in the box provided on the following page.

Heights	Frequency

1 Calculate the mean height of the class.

2 Calculate the mode height for the class.

3 Calculate the median height for the class.

4 Put these heights into a frequency distribution table (with 3 cm intervals).

5 Graph this information as a frequency histogram, using Figure 1.2. Remember to label both axes.

Figure 1.2 Results

6 Provide a written description of the results you have obtained.

PART B

1 Create a scatter plot of the following data (using Figure 1.3), in Table 1.11, collected during an investigation into the relationship between time spent revising and exam scores.

Table 1.11 Relationship between time spent revising and exam scores

Participant	Exam score (%)	Revision time (minutes)
1	94	80
2	88	60
3	71	56
4	75	47
5	71	43
6	68	40
7	73	67
8	57	16
9	59	21
10	65	54
11	67	40
12	60	39
13	60	38
14	45	8
15	61	34
16	50	16

Figure 1.3 Results

2 Explain the results you have obtained.

_____

_____

_____

_____

1.4.2 Reproducibility in research

Key science skills

Analyse and evaluate data and investigation methods
- identify and analyse experimental data qualitatively, applying where appropriate concepts of: accuracy, precision, repeatability, reproducibility and validity; errors; and certainty in data, including effects of sample size on the quality of data obtained

Construct evidence-based arguments and draw conclusions
- identify, describe and explain the limitations of conclusions, including identification of further evidence required

Develop

Read the research article and answer the questions that follow.

How a peanut butter test may detect Alzheimer's

Research on better diagnosis and treatment continues
Cleveland Clinic
15 December, 2020

Creamy or crunchy – and oh, so spreadable – peanut butter is not your first thought as a possible game-changer in Alzheimer's disease research.

But it has potential, according to researchers at the University of Florida. They conducted a peanut butter smell test hoping to find an inexpensive, non-invasive way to detect early-stage Alzheimer's and track its progress.

The test was conducted on cognitively normal individuals as well as:
- 18 patients diagnosed with probable Alzheimer's disease.
- 24 patients with mild cognitive impairment.
- 26 patients with other causes of dementia.

The researchers found that the peanut butter test singled out those with probable Alzheimer's.

Here's how they conducted the test. The researchers asked each person to close their eyes, their mouth and one nostril. They opened a small container of peanut butter and moved progressively closer until the person could smell it. After measuring that distance, they waited 90 seconds and repeated the process with the other nostril.

In those with probable Alzheimer's disease, the researchers had to move the peanut butter container an average of 10 centimetres closer to the left nostril than to the right nostril.

The investigators, who published their study in 2013, said follow up research would be needed.

However, a follow-up study in 2014 at the University of Pennsylvania could not replicate their results. The second research team found no difference in the ability of 15 patients with Alzheimer's to smell peanut butter in their left versus their right nostrils.

'This highlights the scientific importance of studies being repeated and refined by other researchers in different patient populations,' says Dr. Wint. 'Intriguing results don't always hold true across all study populations.'

Research continues on Alzheimer's disease as well as on mild cognitive impairment.

'The accessibility of current Alzheimer's tests is one of the issues that is making diagnosis and research difficult,' notes Dr. Wint.

Currently, the most accurate early-stage diagnostic tests for Alzheimer's are a spinal tap or an amyloid PET scan. However, these tests are expensive, uncomfortable and not available everywhere.

Accurate, accessible and inexpensive testing could inform more patients about their Alzheimer's disease status. And diagnosing Alzheimer's in its early stages is critical to finding treatments that can delay or prevent future memory loss.

Source: adapted from Cleveland Clinic (2020, December 15). How a Peanut Butter Test May Detect Alzheimer's https://health. clevelandclinic.org/2015/10/peanut-butter-test-may-detect-alzheimers

---

1 In the original study what was the population of research interest?

_____

2 What was the control group?

_____

3 What were the experimental groups?

4 Out of the experimental groups, which groups were used as a comparison?

5 What did this research conclude?

6 What is reproducibility in research?

7 Why is it important for research to be reproduced? Use the article to justify your response.

1.4.3 Errors

Key science skills
Analyse and evaluate data and investigation methods
• evaluate investigation methods and possible sources of error or uncertainty, and suggest improvements to increase validity and to reduce uncertainty

Develop

PART A

In Table 1.12, define each type of error that can occur during an investigation.

Table 1.12 Types of error

Error	Definition
Random error	
Systematic error	
Personal error	
Experimenter bias (a type of systematic error)	

Unsurprisingly errors occur quite often.

1 What might indicate to you that you have made an error in your investigation?

2 If you know that you have made a personal error, what should you do?

3 If you want to be sure that an error that looks random is actually a random error, what you should do?

4 Outline a technique you can use to reduce the chance of your results being biased.

PART B

For each investigation listed, determine which of the four main types of error is involved.

1 Orlando wanted to investigate the effect of sleep and caffeine consumption on test performance. He allocated his sample into students who drank two cups of coffee and those who did not drink any coffee. For each of these conditions, he further allocated his sample by number of hours slept: 3 hours, 6 hours and 9 hours. When giving one of the participants their coffee, Orlando lost count and accidentally gave them a cup with three, rather than two, teaspoons of coffee. Orlando was not aware that he had done this. _____

2 To determine whether childhood participation in sports is related to better self-esteem in adolescents, a representative sample of 100 adolescents with a history of sports participation were given a questionnaire on self-esteem and their scores averaged. Another representative sample of 100 adolescents were also given this questionnaire, but they were selected with no history of sports participation. The instrument used to measure self-esteem was the Oxford Happiness Scale. _____

3 A convenience sample of 20 Year 12 students was tested for concentration 30 minutes prior to eating their lunch and then 30 minutes after they had eaten their lunch. The test required them to read a short passage of text and identify 10 simple grammatical errors as quickly as possible. The time taken to identify all errors was measured in seconds. _____

4 Participants were divided into two groups. One group flew a flying simulator after consuming one standard drink; the other group did not consume any alcohol. Further to that, half the participants in each group were experienced pilots with over 10 years' experience, the other half had only just received their license. Reaction time was measured. The simulator had not been calibrated. _____

5 To determine the effect of texting on driving skill, orange traffic cones were set up in a circuit in a car park. A group of students were tested on the circuit, once while receiving and sending texts, and one without texting. Skill was measured by the number of cones hit while driving each circuit. The experimenter did not change the circuit between tests. _____

6 A psychology teacher wanted to test whether creating a mind map of a topic prior to testing would improve test results. He randomly allocated his students into two groups. One group was required to complete a mind map using a specific process in addition to their usual study techniques. The other group studied as usual. When reviewing the test results the teacher noticed a few data points that were unusual. He went and reviewed the tests and determined that he had added up a few tests incorrectly. _____

1.4.4 Analysing data

Key science skills

Analyse and evaluate data and investigation methods

- process quantitative data using appropriate mathematical relationships and units, including calculations of percentages, percentage change and measures of central tendencies (mean, median, mode), and demonstrate an understanding of standard deviation as a measure of variability
- identify outliers and contradictory or incomplete data

Develop

PART A

Prabh was conducting research on the effect of stress on levels of salivary cortisol (a hormone that is released during stress). Refer to his raw data in Table 1.13 and answer the questions that follow.

The relationship between the IV and DV can become clearer after analysing your data.

Table 1.13 Salivary cortisol concentration (n = 50)

Participant	Salivary cortisol (µg/l)	Participant	Salivary cortisol (µg/l)
1	24.2	26	23.9
2	25.3	27	29.2
3	22.0	28	43.5
4	26.4	29	29.0
5	24.5	30	31.1
6	26.2	31	22.2
7	21.0	32	35.4
8	24.2	33	26.3
9	27.2	34	74.2
10	23.9	35	23.9
11	24.3	36	24.2
12	21.0	37	21.5
13	23.6	38	25.6
14	24.9	39	24.8
15	56.2	40	27.4
16	22.9	41	21.8
17	23.8	42	24.5
18	28.5	43	24.4
19	27.4	44	26.8
20	24.2	45	29.1
21	24.2	46	24.7
22	24.1	47	23.1
23	23.6	48	62.3
24	19.2	49	24.1
25	26.4	50	25.8

1 Calculate the mean salivary cortisol concentration from this set of data.

2 Calculate the range for the data.

3 Identify and list any outliers in the data.

4 Calculate the mean salivary cortisol concentration and the range without the outliers. How did this affect each descriptor?

5 How does the inclusion of the outliers in the data affect the standard deviation?

_____

6 How can you determine whether these outliers are due to a random error?

_____

_____

7 Prabh did some further research and he found that cortisol levels change naturally over the course of the day. Why would this be important for Prabh to know prior to completing his measurements?

_____

_____

_____

_____

PART B

Kate wanted to investigate the effect of number of hours study each day, completed the week before a test, on one topic in psychology. She organised for 25 students to study an assigned number of hours each day and collated their test results. These results are included in Table 1.14. Refer to the table and answer the questions that follow.

Table 1.14 Effect of hours of study on test scores

Number of hours of study	Average test scores out of 50 (n = 25)
0	9
1	21
2	32
3	40
4	46

1 Draw a line graph of the data.

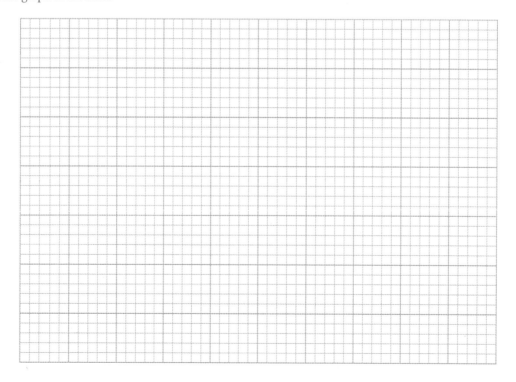

2 Describe the graph.

_____

_____

_____

3 Kate was concerned that the data did not follow a straight-line trend. Account for this result.

_____

_____

_____

4 How could Kate test that the trend was other than a straight line?

_____

_____

PART C

Alzheimer's is a neurodegenerative disease characterised by high levels of a protein, called tau, that accumulates in brain neurons and causes damage. There has been a lot of research to determine whether there are biomarkers found in plasma, urine and saliva that could serve as a diagnostic or monitoring tool given their relative cost effectiveness compared to other forms of measurement.

Below is a set of data from three laboratory assistants in a laboratory testing for tau protein levels in blood plasma (the liquid part of blood) using a new machine (Machine B). Decide how accurate and precise each set of results is, comparing it to results from a machine that has shown over many trials to produce accurate results (Machine A). Use rating words such as high, medium or low.

Table 1.15 Sample measurements of tau protein using two machines

| Assistant | Tau protein reading (pg/mL) | | | | Analysis | |
| | Machine B | | | Machine A | | |
	Measurement 1	Measurement 2	Measurement 3		Accuracy	Precision
1	3.92	3.93	3.90	3.93		
2	4.01	4.01	4.01	3.90		
3	3.33	3.32	3.30	3.88		

Source: adapted from Pase MP, Beiser AS, Himali JJ, et al. (2019). Assessment of plasma total tau level as a predictive biomarker for dementia and related endophenotypes. _JAMA Neurol_, 76(5), 598–606.

1 Which set of measurements is the most precise?

_____

2 Which set of measurements are the most accurate but not precise?

_____

3 What is the best way to improve precision?

_____

_____

4 Assuming we know the true value, what is the best way to improve accuracy?

_____

_____

_____

PART D

Fill in Table 1.17 with the correct type of relationship (positive correlation, negative correlation or uncorrelated) and indicate which scatter plot from Table 1.16 would best fit the data (A, B or C).

Table 1.16 Types of correlation

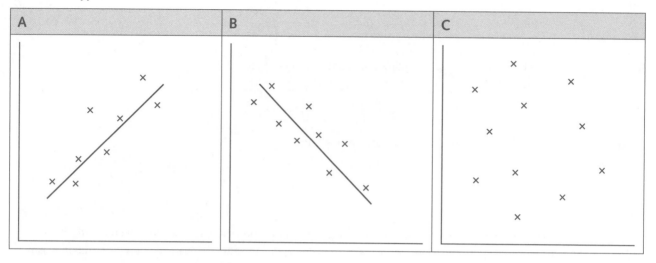

Table 1.17 Relationships between variables

Example	Type of relationship	Scatter plot (A, B or C)
Dr Fitzgerald found that the more babies are held, the less they cry.		
A study was done that found that students who got more sleep received lower exam results.		
Dr Roan found that there is no relationship between coffee consumption and intelligence.		
Dr Gregory examined the relationship between drinking coffee and reaction time, and found that the more coffee a person drank, the faster their reaction time (in seconds).		

1.4.5 Evaluation of research

Key science skills

Develop aims and questions, formulate hypotheses and make predictions
- identify independent, dependent, and controlled variables in controlled experiments
- formulate hypotheses to focus investigation

Analyse and evaluate data and investigation methods
- identify and analyse experimental data qualitatively, applying where appropriate concepts of: accuracy, precision, repeatability, reproducibility and validity; errors; and certainty in data, including effects of sample size on the quality of data obtained

Construct evidence-based arguments and draw conclusions
- identify, describe and explain the limitations of conclusions, including identification of further evidence required

Develop

Read the activity scenario and then answer the questions that follow.

A psychologist wanted to test whether or not information could be consolidated while an individual was sleeping. She advertised in a local newspaper for participants aged between 18 and 20 years and sampled 100 applicants by drawing their names out of a hat. The participants were exposed to two conditions.

The control condition involved participants reading a list of 50 words when they woke up after a night's sleep. Before going to bed the following evening, they were asked to write down as many words as they could remember from the list. The same participants were then used in the experimental condition, which involved participants reading a similar list of 50 words before going to bed and then writing down as many as they could recall upon waking from sleep.

The results showed that, during the control condition, participants remembered an average of 33 per cent of the words. During the experimental condition, participants remembered an average of 62 per cent of the words. These results were found to be statistically significant.

1   Is this study an experiment? Explain your answer.

_____

_____

_____

2   Write a suggested hypothesis for this study.

_____

_____

3   What is the independent variable for this study?

_____

4   What is the dependent variable for this study?

_____

5   What sort of sampling procedure was used in this experiment?

_____

6   What is one advantage of using this sort of sampling procedure?

_____

7   What is an alternative sampling technique that could be used?

_____

_____

8   Explain how a sample could be obtained using this technique.

_____

_____

_____

9   Write a conclusion for this study.

_____

_____

_____

1.4.6 The ethics spider

Key science skills
Analyse, evaluate and communicate scientific ideas
- analyse and evaluate psychological issues using relevant ethical concepts and principles, including the influence of social, economic, legal and political factors relevant to the selected issue

Develop

MATERIALS

- coloured highlighters

INSTRUCTIONS

1   Read the activity scenario and use one colour to highlight where ethics are adhered to and another colour to highlight where ethics are breached.

A new football boot known as Powerdrive was about to hit the market. The boot was intended to increase kicking distance.

The directors at Powerdrive wanted to test the boot in trials at the MCG. They contacted each AFL club and asked them to find two players who would volunteer to be involved in a study, but they did not explain what the study was. When the players arrived at the MCG, the directors did not tell them exactly what was going to happen because they didn't want their expectations to affect the results; however, they let the players know that if they felt uncomfortable at any time, they could cease their involvement in the study.

First, they recorded each player's average kicking distance when wearing their own boot, then their average kicking distance wearing the Powerdrive boot. The players were then told what the results showed and why they had conducted the comparison between the two different types of boots. The results showed an average improvement in kicking distance of 1.16 m when wearing the Powerdrive boot compared to their regular boots.

The directors were thrilled with the results and published each player's name and kicking distance in the newspaper to launch their campaign.

2   Figure 1.4 shows eight ethical considerations. Define each of these in the appropriate place in the figure.

3   In the 'Where?' boxes, explain where each ethical consideration occurred in the scenario, and why you believe it was breached or adhered to.

Figure 1.4 Ethics spider

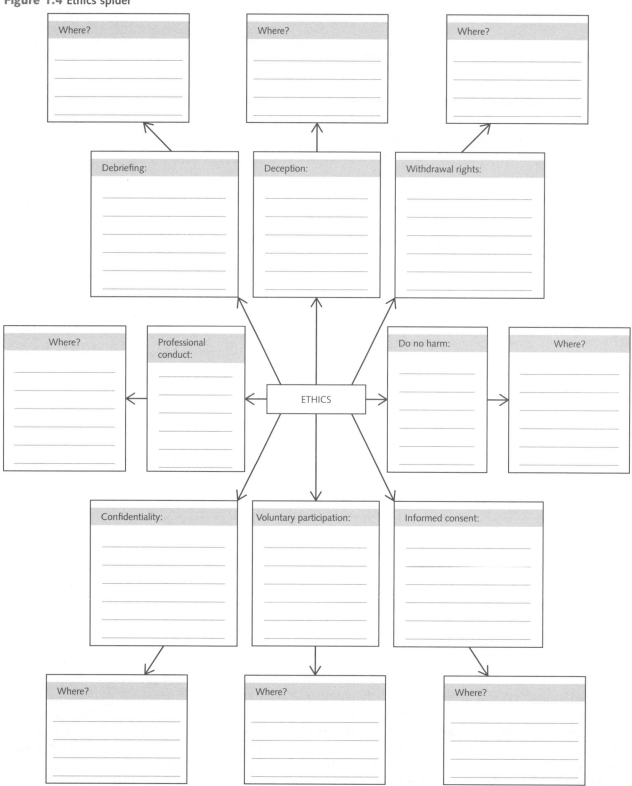

1.4.7 Ethics

Key science skills
Analyse, evaluate and communicate scientific ideas
* analyse and evaluate psychological issues using relevant ethical concepts and principles, including the influence of social, economic, legal and political factors relevant to the selected issue

Develop

PART A

Define each of the following ethical guidelines and provide an example of how each can be breached.

Table 1.18 Ethical guidelines

Ethical guideline	Definition	Example of breach
Justice		
Integrity		
Maleficence		
Nonmaleficence		
Respect		

PART B

Identify the ethical guideline represented by each of the following scenarios.

1 A researcher has been brought in to join a team of researchers investigating brain trauma. One of the instruments that is used to measure brain trauma is a PET scan. The doctor in charge of the laboratory outlines a procedure for setting up the machine that is not in the manual and appears to bias the analysis. The researcher completes the procedure as instructed but then decides to repeat the imaging using the procedure outlined in the manual. There is a difference in the results. The doctor in charge ignores the second set of results. _____

2 A researcher continues to use brain surgery to isolate areas of brain function despite there being less invasive techniques available. _____

3 A new drug has been developed to treat schizophrenia, a debilitating mental disorder; however, it is extremely costly so is only available for patients who have private health insurance or are relatively wealthy. _____

4 During an observational study on maternal care practices of an Aboriginal community, a researcher films the group, causing distress. _____

5 Participants gave informed consent to take part in an investigation studying the perception of violence. They were shown graphic images that were upsetting. The researchers failed to provide an opportunity for participants to seek counselling, if required. _____

Chapter 1 summary

PART A

Multiple-choice questions

Circle the correct answers.

1 The usual reason why an experimenter uses deception in research is because:

   A the experiment will cause some distress and the experimenter does not want the participant to withdraw.

   B the participant is a university student studying psychology and is therefore not naïve.

   C the research is testing morals and it could be embarrassing or upsetting.

   D if participants have too much information, they will change their way of behaving and influence the results.

2 Researchers try to use random samples in their studies because:

   A they want to select a sample that is representative of the population.

   B it enables them to eliminate any special characteristics that the participants may possess.

   C it helps them eliminate any experimenter effects.

   D most statistics are based on the assumption that the samples are randomly selected.

3 Of the steps listed below, which step occurs at the latest stage of psychological research?

   A Identify the research problem

   B Formulate a hypothesis

   C Design the method

   D Report the research findings

4 A researcher was interested in testing whether playing violent video games leads to an increase in aggressive behaviours in children. In this study, the independent variable and the dependent variable are, respectively:

   A number of aggressive behaviours; age of the child.

   B playing violent video games or not; age of the child.

   C number of aggressive behaviours; playing violent video games or not.

   D playing violent video games or not; number of aggressive behaviours.

5 Sloan and his colleagues (1975) wanted to test the effectiveness of different forms of psychotherapy. Ninety people, who all had the same personality disorder, were put into three groups. The first group (Group 1) received counselling, the second group (Group 2) were given behaviour therapy and the third group (Group 3) were put on a waiting list. After four months, 80 per cent of participants in Groups 1 and 2 showed improvement. By comparison, 48 per cent of the people in Group 3 showed improvement. The control group in this study was:

   A the group receiving counselling.

   B the group receiving behaviour therapy.

   C the group on the waiting list.

   D none of the three groups – there is no control group.

Questions 6–9 relate to the following information.

In an experiment investigating how sleep deprivation affects memory, one group of participants were kept awake for three days and nights and another group was told to sleep as they normally would. At the end of that time, all participants were given a learning task and were tested on their recall of the learned material.

6 In this experiment, the independent variable and the dependent variable are, respectively:

   A number of hours of sleep deprivation; score on the recall test.

   B score on the recall test; number of hours of sleep deprivation.

   C presence of two groups; learning task.

   D learning task; presence of two groups.

7 The experimenters conducting the study ensured that all participants experienced the same temperature and light conditions during the test period. In this experiment, what kind of variables were light and temperature conditions?

A Independent

B Dependent

C Extraneous

D Unwanted

8 The group that was sleep deprived for three days and nights was the:

A experimental group.

B control group.

C independent group.

D dependent group.

9 The most appropriate hypothesis for this experiment would be:

A 'Participants who are sleep deprived would perform just as well on the learning test as participants who are not sleep deprived'.

B 'It will be investigated whether sleep deprivation decreases the ability to remember learnt tasks'.

C 'Participants who are sleep deprived will perform better on the learning task compared to participants who are not sleep deprived'.

D 'Participants who are sleep deprived will perform worse on the learning task compared to participants who are not sleep deprived'.

10 In psychological research, reliability refers to:

A whether you can rely on your participants in an experiment.

B whether you can rely on the experimenter in an experiment.

C an experiment consistently measuring what it is supposed to measure.

D an experiment measuring what it is supposed to measure.

Questions 11–14 relate to the following information.

An experimenter randomly allocated 50 first-year university maths students to one of two math tutorials, to be taught by the same tutor. The tutor was told that the students were placed into the groups according to their mathematical ability. The students were not told that they were taking part in a research activity. After three months, all students sat the same exam. It was found that the students who had been labelled as 'high ability' performed better on the exam than those labelled 'low ability'.

11 The most likely explanation of these findings is that:

A the students in the 'high ability' group really were more intelligent.

B students' expectations influenced the outcome.

C the outcome was influenced by the experimenter's expectations of the participants.

D the results were biased because of the placebo effect.

12 The study was criticised for failing to satisfy ethical considerations because the experimenter:

A did not debrief the participants at the completion of the study.

B allowed students to experience physical harm.

C allowed students to experience psychological harm.

D revealed the name of the students to the tutor and thus did not maintain confidentiality.

13 In this experiment, the sample and the population were, respectively:

A low-ability maths students; high-ability maths students.

B high-ability maths students; low-ability maths students.

C university students; 50 first-year university maths students.

D 50 first-year university maths students; university students.

14 The experimenter used random allocation to assign participants to groups. Which of the following procedures was he likely to follow?

A Line up all 50 students in height order, then place the shorter 25 students in Group 1 and the taller 25 students in the Group 2.

B Place all males in the one group and all females in another group.

C Ask students which group they want to be in and allocate accordingly.

D Put the names of all 50 participants in a hat; draw out 25 names for Group 1 and place the remaining 25 names in Group 2.

15 Which of the following does not belong in a method section of a scientific report?

A Participants

B Hypothesis

C Materials

D Procedure

16 'Random allocation' means:

A each participant is selected from a group that is randomly selected and therefore representative of the population.

B each participant has an equal chance of being assigned to either the experimental or control group.

C each participant is chosen at random from the population to be in the sample.

D the groups that are formed differ with respect to some critical variable.

17 Table 1.19 shows the age distribution of adults at the Elite Fitness Sports Club, which has more than 1000 members at a number of venues.

Table 1.19 Age distribution of members of the Elite Fitness Sports Club

Age range (years)	% in age category
16–30	40%
31–45	20%
46–60	25%
61–75	15%

A stratified sample of 100 individuals is drawn from this population. Which alternative best describes the composition of this sample?

A 60 people aged 16–30, 20 people aged 31–45, 30 people aged 46–60 years, 15 people aged 61–75

B 40 people aged 16–30, 20 people aged 31–45, 25 people aged 46–60 years, 15 people aged 61–75

C 25 people aged 16–30, 25 people aged 31–45, 25 people aged 46–60, 25 people aged 61–75

D There is insufficient information to answer this question.

18 Which of the following statements best describes why stratified sampling is used?

A Stratified sampling is used to select a sample that is representative of the population.

B Stratified sampling is used to control participant-related extraneous variables.

C Stratified sampling is used to ensure correct proportions of participant characteristics are represented in the sample.

D All of the above

Questions 19–20 relate to the following information.

Lidia is a Year 11 psychology student writing her first scientific research report. She has carried out an experiment on help-seeking behaviour and how it differs depending on whether a person is alone or in a group.

19 Lidia needs to describe past research in the area of help-seeking behaviour. In what section of the scientific report should she include this information?

A Introduction

B Method

C Discussion

D References

20 Lidia hypothesised that 'people are more likely to help others in need if they are alone rather than in a group'. In what section of the scientific report does she need to include her hypothesis?

A Introduction

B Method

C Discussion

D References

PART B

Complete the crossword below.

Figure 1.5 Summary crossword

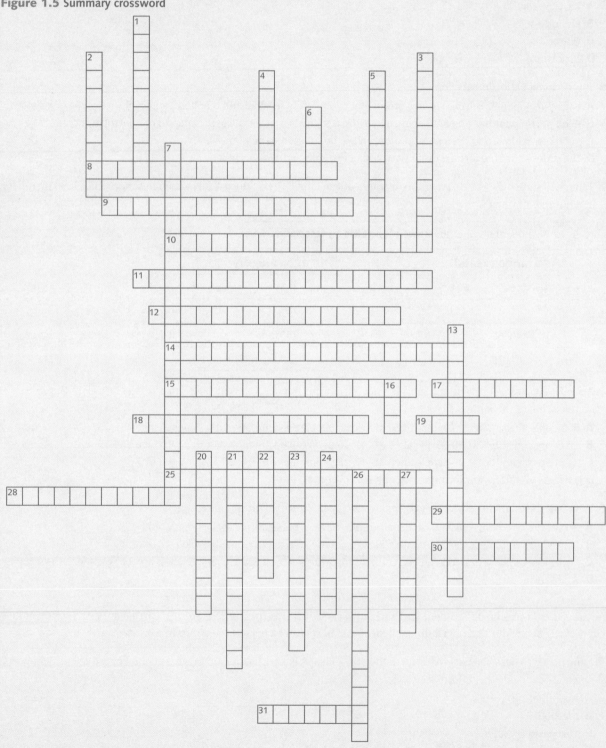

Across

8 A method used by researchers to control order effects by systematically exposing different participants to different orders of treatment conditions

9 This occurs when the results can be generalised to populations beyond the investigation sample

10 A measure of the variability of a set of scores or values within a group, indicating how narrowly or broadly they deviate from the mean

11 A variable that has affected the results (dependent variable), apart from the independent variable

12 Assigning participants to either the experimental group or control group in an experiment, ensuring that all participants have an equal chance of being allocated to either group

14 The factor or condition (variable) that an experimenter manipulates (changes or varies) systematically to determine its effect on another variable (the dependent variable)

15 Results that are attributable to a researcher's behaviour, preconceived beliefs, expectancies or desires about results

17 The measure that would have been observed on a construct were there no error involved in its measurement

18 Data that records or describes the attitudes, behaviours or experiences of participants conceptually

19 The most frequently occurring score in a set of data

25 A study that collects data over two or more periods in time, using the same participants

28 A research tool that assigns a numbers on a scale to indicate the degree agreement with a statement or the frequency of some behaviour

29 A procedure where participants do not know which experimental condition they have been assigned to, but the experimenter does

30 A form of research investigation in which the researcher observes and interacts with a selected environment beyond the laboratory

31 Any medical or psychological treatment that is known not to affect the condition being studied, making it useful as a control condition to compare with the effects of a treatment of interest

Down

1 How closely a set of measurement values are to one another; determined by the repeatability and/or the reproducibility of the measurements obtained using a particular measurement instrument and procedure

2 How close a measurement is to the 'true' value of the quantity being measured. Accuracy is not quantifiable; measurement values may be described as more accurate or less accurate

3 A measure of the strength of relationship between two variables

4 A measure of central tendency, calculated by arranging scores in a data set from the highest to the lowest, and selecting the middle score

5 This relies on participants answering questions honestly that relate to their feelings, beliefs, attitudes or behaviours

6 A measure of central tendency that gives the numerical average of a set of scores, calculated by adding all the scores in a data set and then dividing the total by the number of scores in the set

7 A randomly selected group of participants that accurately reflects the characteristics of the larger population from which it is drawn

13 The closeness of the agreement between the results of measurements of the same quantity being measured, carried out under changed conditions of measurement

16 The purpose of an investigation

20 The commitment to searching for knowledge and understanding, the honest reporting of all sources of information and results

21 A design where the same group is both the control or comparison group as well as the group exposed to the independent variable

22 Data points within a data set that are distant from the majority of other values in the data set

23 A research design that includes both within and between subjects conditions as independent variables

24 A research investigation that focuses on a particular person, activity, behaviour, event or problem that is, or could be, experienced within a real-world context outside of the laboratory

26 This occurs when measurements are consistently different from the true value each time the measurement is made

27 When participants are deliberately misled or not fully informed of the true nature or purpose of a research investigation

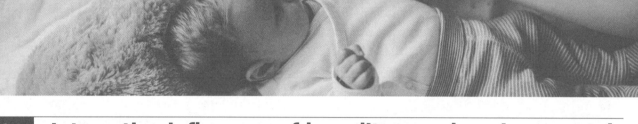

2 Psychological development

Interactive influences of hereditary and environmental factors

Key knowledge
- the interactive influences of hereditary and environmental factors on a person's psychological development
- the biopsychosocial approach as a model for considering psychological development and mental wellbeing
- the process of psychological development (emotional, cognitive and social development) over the course of the life span
- the role of sensitive and critical periods in a person's psychological development

2.1.1 Environment and infant development

Key science skills
Develop aims and questions, formulate hypotheses and make predictions
- identify independent, dependent, and controlled variables in controlled experiments
Construct evidence-based arguments and draw conclusions
- discuss the implications of research findings and proposals, including appropriateness and application of data to different cultural groups and cultural biases in data and conclusions

Develop

Read the case study and answer the questions that follow.

CASE STUDY

To test the effect of environmental experiences on the motor and cognitive development of babies in their first 12 months, researchers designed an experiment using two groups of babies. They devised a range of tests to determine the babies' levels of motor and cognitive development before the experiment began. They proposed that the first group (Group A) would be exposed to an 'enriched' environment for 9 hours a day, five times a week, for a six-month period. The second group (Group B) would experience minimal enrichment for the same period of time. At the end of this period, both groups would be re-tested using the same tests they had initially been given.

The researchers approached a local day-care centre and explained to the manager what they were interested in doing with the children. They asked the manager if she would select 12 healthy infants in her care, ranging in age from 6 to 12 months, for them to study. The researchers stressed that no physical harm would come to the babies.

Their proposal was that Group A (three boys and three girls) would spend their days in an enriched environment where classical music and nursery rhymes in a variety of languages would be played; the wallpaper would be bright and have interesting pictures; and the room would have a variety of hanging mobiles (which would be changed regularly). This group could play with a variety of toys, which would be regularly exchanged for others. These infants could spend as much time as possible outside of their cots. Twice a day they would be taken outside to explore the garden. Six child-care workers (whose wages would be provided by the researchers) would always be with them and they would read to the children, play with them, talk to them and regularly cuddle them.

Group B (three boys and three girls) would spend their days in a room where the white walls were bare, no music would be played, no mobiles would hang from the ceiling, the toys available would not be changed and the infants would be confined to their cots for 6 hours a day. They would not be taken outside and their interaction with their caregivers would be limited to nappy changes, feeding and the act of being placed into or taken from their cots.

The manager agreed to the researchers' proposal and the experiment began the following week.

Questions

1 Write an aim for the experiment.

2 Identify the independent and dependent variables.

   a Independent variable

   b Dependent variable

3 Write a research hypothesis for the experiment.

4 Identify the sampling procedure and the participants in the experiment.

5 Identify and explain three ethical guidelines the researchers failed to follow.

6 Suggest an alternative way to test the impact of an enriched or impoverished environment on infant development.

2.1.2 Nature vs nurture

Key science skills
Analyse, evaluate and communicate psychological ideas
- analyse and explain how models and theories are used to organise and understand observed phenomena and concepts related to psychology, identifying limitations of selected models/theories

Develop

PART A

Decide on a particular characteristic or ability that you have; for example, a skill in cooking, a sporting ability, artistic or musical strengths or some practical ability. In Table 2.1, brainstorm a list of factors that contribute to this characteristic and write them in the left side of the table under 'Factors' (aim for at least 10 factors). On the right side of the table, identify whether each factor is something that is influenced by hereditary factors, environmental factors or both.

Table 2.1 Factors that affect a characteristic or ability

Factors	Description of the factor	Hereditary/environmental/both
1		
2		
3		
4		
5		
6		
7		
8		
9		
10		

PART B

Time for some fun!

MATERIALS

- coloured markers
- A4 paper

INSTRUCTIONS

1 Using Table 2.1, create a mind map showing how each factor contributes to the development of your chosen characteristic.

2 Draw the mind map by writing the characteristic in the centre of a page of A4 paper. Write each factor that contributes to this characteristic around the central characteristic, with those factors that are more influential closer to the centre and those that are not as influential further from the centre.

3 Draw lines that connect the factors and the characteristic and label each line to show how each factor influences the others.

4 Draw a coloured box around each of the hereditary factors (blue), environmental factors (green) and those that are influenced by both (red).

Figure 2.1 shows an example of a mind map for factors influencing the ability to do a ballet pirouette.

Figure 2.1 Factors affecting the ability to do a ballet pirouette

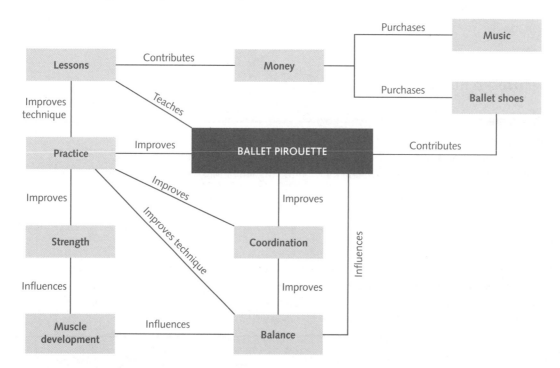

2.1.3 Twin and adoption studies

Key science skills

Plan and conduct investigations

- design and conduct investigations; select and use methods appropriate to the investigation, including consideration of sampling technique (random and stratified) and size to achieve representativeness, and consideration of equipment and procedures, taking into account potential sources of error and uncertainty; determine the type and amount of qualitative and/or quantitative data to be generated or collated

Develop

Complete Table 2.2 by listing at least two advantages and two disadvantages of each type of study for illustrating the influences of heredity and environment on development.

Table 2.2 Some advantages and disadvantages of twin and adoption studies

Twin studies	Adoption studies
Advantages:	Advantages:
Disadvantages:	Disadvantages:

2.1.4 Indigenous games used for motor development

Key science skills

Analyse, evaluate and communicate scientific ideas

- analyse and evaluate psychological issues using relevant ethical concepts and principles, including the influence of social, economic, legal and political factors relevant to the selected issue

Develop

In Table 2.3, list the names of games that are traditionally played in primary school or secondary school sports classes. Using the resource, *Yulunga: Traditional Indigenous Games*, rename the games you have listed and provide alternative equipment that are culturally appropriate for Indigenous students.

Table 2.3 Inclusive teaching strategies

Traditional	Indigenous alternative

2.2   The biopsychosocial model

Key knowledge
- the biopsychosocial approach as a model for considering psychological development and mental wellbeing

2.2.1 Biopsychosocial approach

Key science skills
Construct evidence-based arguments and draw conclusions
- analyse and explain how models and theories are used to organise and understand observed phenomena and concepts related to psychology, identifying limitations of selected models/theories

Develop

Study the contributing factors listed in Table 2.4. Write each one into Figure 2.2 under either the biological, psychological or social section of the biopsychosocial model of health and illness.

Table 2.4 Contributing biological, psychological and social factors

stress	genetic vulnerability	socio-economic status	attitudes
loss of a significant relationship	memory	emotional regulation	coping strategies
poor sleep	nutrition	hormones	bullying
culture	media	friendships	parenting
personality	learning		

Figure 2.2 Biopsychosocial model of health and illness

Biological factors:

Psychological factors:

Social factors:

Be sure to add the factors in the right places.

2.2.2 Application of the biopsychosocial model

Key science skills
Analyse, evaluate and communicate scientific ideas
- analyse and explain how models and theories are used to organise and understand observed phenomena and concepts related to psychology, identifying limitations of selected models/theories

Develop

Read the passage below and complete the task that follows.

Experiencing traumatic events is unfortunately commonplace and, in some cases, may lead to the onset of debilitating mental health disorders, such as post-traumatic stress disorder (PTSD).

Many people who have been diagnosed with PTSD also present with depression and anxiety. This can make it difficult to diagnose accurately and create an effective treatment plan. Researchers are investigating the presence of biomarkers present in PTSD individuals that are not found in depressed or anxious individuals. A recent study (2020), using 20 individuals with PTSD and 20 control individuals, aimed to identify differences in blood biomarkers able to distinguish PTSD participants from controls. Significant differences were found in levels of growth hormone, cholesterol and inflammatory molecules between individuals with PTSD and controls.

A second study (2020) found impaired cortisone regulation due to the presence of another molecule in individuals with PTSD compared with controls, subjects exposed to trauma without PTSD and patients with major depressive disorder (MDD).

Biomarkers could be used in combination with psychological criteria, in a biopsychosocial model, to support clinical management decisions and ensure appropriate individual treatment pathways.

Using the internet, research the biological, social and psychological factors that influence PTSD and add these to Table 2.5.

Table 2.5 PTSD: application of a biopsychosocial model

Biological	Psychological	Social

2.3   Psychological development over the life span

Key knowledge
- the process of psychological development (emotional, cognitive and social development) over the course of the life span

2.3.1 Attachment

Key science skills
Develop aims and questions, formulate hypotheses and make predictions
- identify independent, dependent, and controlled variables in controlled experiments
Analyse, evaluate and communicate scientific ideas
- analyse and explain how models and theories are used to organise and understand observed phenomena and concepts related to psychology, identifying limitations of selected models/theories

Develop

PART A

Read the case study and answer the questions that follow.

CASE STUDY

To test and measure the quality of an infant's attachment to their primary caregiver, Ainsworth (1972) created a laboratory experiment known as the 'strange situation'. The mother and infant would enter a room where the infant was free to play and explore (toys were provided), before a stranger would enter the room and sit down. The mother would then leave the room, leaving the infant in the room with the stranger for a 3-minute period. After this time the mother would return to the room. The observer recorded three types of behaviours in the infant: contact-seeking behaviour (child seeks mother or stranger); exploratory play and behaviour; and crying or distress. The quality of the infant's play and the level of distress (crying) when the mother was present was compared to when the infant was left in the room with a stranger and the mother was absent.

Ainsworth was able to develop three categories of infant behaviour from these observations. The first category, Group A, is called the avoidant attachment type, as infants seemed relatively indifferent to their mothers. They rarely cried when their mother left them alone in the room and showed little attention to her when she returned.

Group B were the securely attached type; they protested vigorously when their mothers left the room, they anxiously looked for her during her absence, and responded with pleasure and gave her a lively greeting when she returned. Group A's exploration of the room was not affected by the absence of their mother, whereas Group B explored more when their mother was in the room.

Group C infants, who were the anxiously attached type, showed great distress when their mothers left the room, and were not calmed easily when she returned. Group C infants also appeared to be more anxious than any of the other groups before, during and after the separation from their mothers, and they did not actively explore the room even when their mother was in the room.

The 'strange situation' seems to be a good measure of the attachment of infants to their mothers for societies in countries such as Australia, New Zealand, Canada and the United States. The scenario is not unfamiliar to infants in these societies, because it is fairly normal for infants to be left in the care of babysitters (family or child care) or on the lounge room floor to play with toys. In other societies, it is not the norm to leave infants with babysitters or alone on the floor. Instead, it is normal that an infant would be carried most of the time and cared for primarily by their mother.

The results of placing culturally different infants in the strange situation are completely different. For example, in a study of Japanese infants, more than 40 per cent of infants were classified as being Group C type babies, or insecurely attached to their mothers. This was because the infants became very upset when their mother left the room and refused to be comforted when she returned (Miyake, Chen & Campos, 1985).

Even with these limitations, the strange situation has provided a wealth of information about the processes involved when infants become attached to their caregivers (Peterson, 1989).

Questions

1 What was the aim of the experiment?

2 Identify the participants in the experiment.

3 What was the method used by Ainsworth?

4 What were the results of the experiment?

5 What are the conclusions from the experiment?

6 Explain why the experiment was unsuccessful in some cultures.

Use the clues provided to complete the crossword in Figure 2.3.

Figure 2.3 Crossword of attachment terms

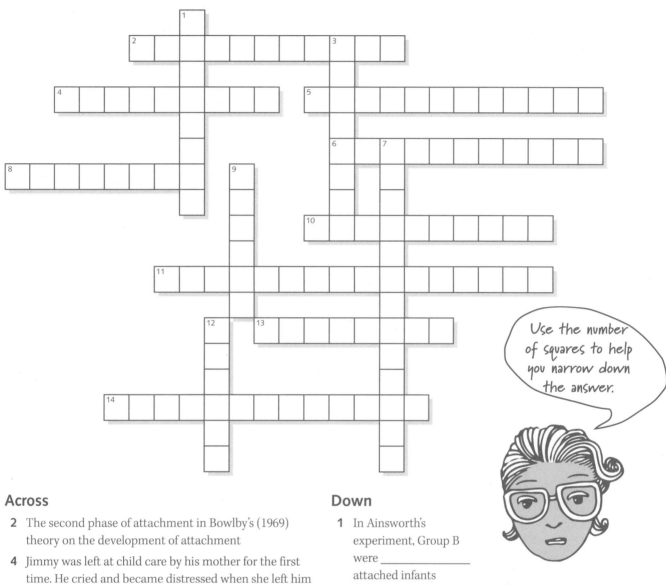

Use the number of squares to help you narrow down the answer.

Across

2 The second phase of attachment in Bowlby's (1969) theory on the development of attachment

4 Jimmy was left at child care by his mother for the first time. He cried and became distressed when she left him with the carer. At 5 p.m. when she returned, he cried when he saw her and was not easily calmed. According to Ainsworth, he is what type of attached child?

5 The final phase of attachment according to Bowlby: the _____ partnership

6 The loss or withholding of important attention and care

8 Attachment theorists believe that attachment has a biological basis, because its main function is to increase chances of _____ for an infant

10 Mary Ainsworth worked in the area of infant–mother _____

11 The name for the scenario Ainsworth developed to test the bond between an infant and its mother (7,9)

13 According to Bowlby, when an infant will move around the environment, using their caregiver as a secure base, they are in this phase of attachment; phase 3 (5-3)

14 The name given to children found living in poor conditions (5,8)

Down

1 In Ainsworth's experiment, Group B were _____ attached infants

3 In Ainsworth's experiment, Group A were _____ attached infants

7 According to Bowlby, when an infant responds to people but does not discriminate between them, they are in this phase of development

9 Infants engage in these behaviours to gain attention from their mothers

12 British psychiatrist John _____ believed that children who suffered loss and failure in early relationships were more likely to have negative effects in later life

2.3.2 Harlow's monkeys

Key science skills

Analyse, evaluate and communicate scientific ideas
* analyse and explain how models and theories are used to organise and understand observed phenomena and concepts related to psychology, identifying limitations of selected models/theories

Develop

Complete the flowchart in Figure 2.4 to outline Harlow's experiment on attachment.

Figure 2.4 Harlow's experiment on attachment in monkeys

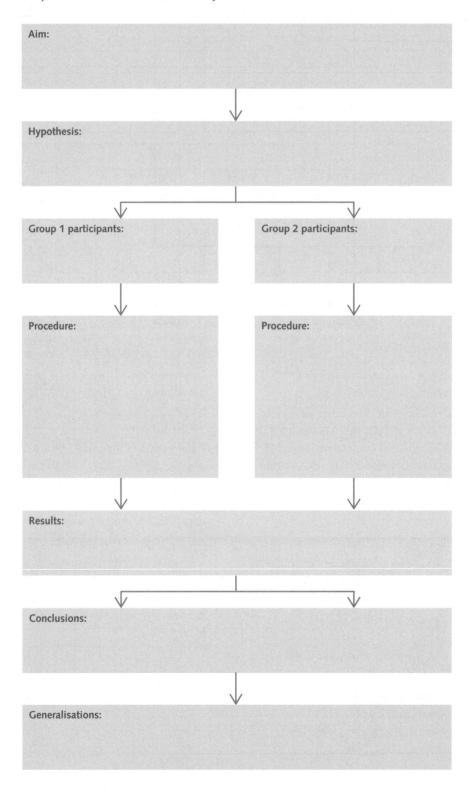

2.3.3 Effect of deprivation: data analysis

Key science skills

Analyse and evaluate data and investigation methods
- process quantitative data using appropriate mathematical relationships and units, including calculations of percentages, percentage change and measures of central tendencies (mean, median, mode), and demonstrate an understanding of standard deviation as a measure of variability

Develop

The information below summarises a study involving early childhood deprivation. Examine the details and answer the questions that follow.

Early childhood deprivation is associated with alterations in adult brain structure despite subsequent environmental enrichment

Early childhood deprivation is associated with higher rates of neurodevelopmental and mental disorders in adulthood. The impact of childhood deprivation on the adult brain and the extent to which structural changes affect the brain are currently unknown. To investigate these questions, researchers utilised MRI data collected from young adults who were exposed to severe deprivation in early childhood in the Romanian orphanages of the Ceauşescu era and then subsequently adopted by UK families. The sample consisted of 67 Romanian adoptees (with between three and 41 months of deprivation) who were compared with 21 non deprived UK adoptees. Table 2.6 shows the data collected.

Table 2.6 Characteristics measured in the study

Measure	Condition	Mean
Young adult body height (cm)	non deprived	177.5
	deprived	164.1
Young adult ADHD symptoms (number)	non deprived	1.2
	deprived	4.1
Young adult IQ	non deprived	106.7
	deprived	95.4
Young adult brain volume (cm³)	non deprived	1190
	deprived	1080

Figure 2.5 Scatter plot of deprivation duration and brain volume

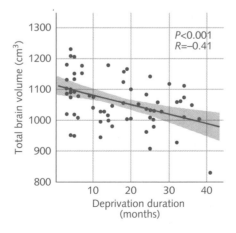

Source: Figure 1B from Mackes, NK., Golm, D., Sakar, S., Kumsta, R., Rutter, M., Fairchild, G., Mehta, M., & Sonuga-Barke, EJS. (2020). Early childhood deprivation is associated with alterations in adult brain structure despite subsequent environmental enrichment. *Proceedings of the National Academy of Sciences*, 177(1), 641– 649.

1 What is the percentage difference in brain volume between groups?

_____

2 What do the results tell you about the number of ADHD symptoms reported by each group?

_____

3 What does Figure 2.5 tell you about brain volume and number of months deprived?

_____

4 Table 2.6 has given mean values for each characteristic. Is this enough to tell you whether the difference is significant? Explain.

_____

_____

_____

_____

_____

5 What other variables besides deprivation could account for these differences?

_____

_____

Scatter plots can help us see trends.

2.3.4 Piaget's theory of cognitive development

Key science skills

Analyse, evaluate and communicate scientific ideas

- analyse and explain how models and theories are used to organise and understand observed phenomena and concepts related to psychology, identifying limitations of selected models/theories

Develop

MATERIALS

- poster cardboard
- scissors
- glue

INSTRUCTIONS

1 Study the definitions in Table 2.7 and the terms in Table 2.8.

2 Cut out the strips of definitions in Table 2.7. (Do not cut each rectangle individually yet.)

3 Glue the strips of definitions to your poster cardboard, then cut out the individual cardboard-backed rectangles of definitions. Be careful not to lose any.

4 Cut out the individual rectangles for the terms in Table 2.8.

5 Match each term to its correct cardboard-backed definition and glue them back-to-back.

6 Use the study cards when revising this topic.

Remember to read the definitions carefully.

Table 2.7 Definitions of terms used in Piaget's theory of cognitive development

A way of thinking that does not need a visual cue in order to understand concepts	When infants can only focus on one quality or feature of an object at a time
Understanding that something can change from one state to another	When an infant learns that an object does not change its weight, mass, volume or area when the object changes its shape or appearance
The use of existing mental patterns in new situations	The ability to use symbols, such as words and pictures, to represent objects, places or events
When infants are incapable of following a line of thought back to its original starting point	The stage of development where children are mainly learning about their environment through senses and purposeful movement
The stage in which children begin to use language and think symbolically	The modification of existing mental patterns to fit new demands
During this stage children begin to use the concepts of time, space and number; they learn that Santa Claus is not real	When infants have difficulty seeing things from another person's perspective
Behaviour that is carried out with a particular purpose in mind	The changes that occur in human thinking, knowing and understanding
Thinking that makes little or no use of reasoning or logic	Understanding that an object still exists even if it cannot be seen or touched
When individuals can develop plans to solve problems and identify a range of possible solutions to a problem	The ability to organise information into categories based on the common features that set them apart from other groups of things
The final stage of cognitive development, characterised by thinking that includes abstract, theoretical and hypothetical ideas	Possibilities based on a supposition, guess or projection

Table 2.8 Terms used in Piaget's theory of cognitive development

Sensorimotor stage	Abstract thinking
Centration	Logical thinking
Hypothetical possibilities	Symbolic thinking
Concrete operational stage	Goal-directed behaviour
Classification	Accommodation
Transformation	Conservation
Reversibility	Formal operations stage
Intuitive	Object permanence
Egocentrism	Preoperational stage
Cognitive development	Assimilation

Table 2.8 Terms used in Piaget's theory of cognitive development

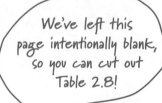

PART B

Use the clues, your knowledge of the development of individual behaviour, your notes and the textbook to complete the crossword in Figure 2.6.

Figure 2.6 Crossword of terms used in Piaget's theory

Across

4 Jamir is nine months old. He is sitting in a highchair near a table. He pulls the tablecloth so that he can reach a banana. According to Piaget, this is an example of what behaviour? (4,8)

5 The type of thinking that makes little or no use of reasoning or logic

7 The final stage of Piaget's theory commences at approximately this age

10 This is the concept that mass, weight and volume remain unchanged when the shape of an object changes

13 Piaget suggested that during the formal operations stage of our development, we begin to think in these terms

14 According to Piaget, in this stage of cognitive development children develop an awareness that items continue to exist even though they cannot see them

15 What stage of cognitive development would Piaget say a child was in if they could solve the following problem? Sarah is taller than Louise. Louise is taller than Richelle. Is Sarah taller or shorter than Richelle? (6,11)

16 This is the third stage of cognitive development, according to Piaget's theory (8,11)

17 Piaget's first stage of cognitive development commences at _____ and concludes at approximately two years of age

18 The second stage of cognitive development, according to Piaget's theory, commences at approximately this age

Use the number of squares to help you narrow down the answer.

19 The progressive changes that occur in human thinking, knowing, understanding, problem-solving and information-processing are known as this kind of development

20 Piaget suggested that children have developed this ability when they can understand new information by fitting it into their existing thinking patterns

Down

1 A four-year-old child who likes the taste of lemons and cannot understand that other people may dislike this taste may be described as this

2 According to this theorist, children's cognitive development is characterised by distinct stages

3 The third stage of cognitive development, according to Piaget's theory, commences at approximately this age

6 According to Piaget, children have developed this mental ability when they realise that something can change from one state to another

8 The type of possibilities based on supposition, guesses or projection

9 I am three years old. I can use language to communicate, but I have trouble understanding situations from other people's points of view. I am in this stage of Piaget's cognitive development

11 Children who believe that a toy elephant no longer exists because it is placed under a pillow have not yet developed this concept (6,10)

12 Piaget suggested that children have developed this when they can change their thinking patterns so that new and different information can be fitted in without this information being changed

PART C

Use the terms in Table 2.9 to fill in the blanks in the following paragraphs. (Note: Some terms may be used more than once.)

Table 2.9 Terms used in Piaget's theory of cognitive development

assimilation	egocentric	age 11 and up	reversibility of thought
seven to 11 years	stages	sensorimotor	accommodation
concepts	birth and two years	conservation	hypothetical possibilities
abstract principles	observations	two and seven years	preoperational
mental frameworks	object permanence	concrete operational	formal operations

Piaget's _____ of children led him to conclude that, generally speaking, children
do not think, reason or problem-solve like adults. Piaget suggested that children's cognitive processes develop gradually,
and that this development occurs in _____. Piaget also suggested that two processes,
which he labelled _____ and _____, underlie this
cognitive development. According to Piaget, _____ refers to a child's tendency to fit
new information into existing _____, so their understanding of the world
is based on their existing _____ and ways of thinking. The
process of _____ involves the tendency for the child to alter their
existing _____ concepts or thought patterns so new information can fit in.

According to Piaget, the first stage in cognitive development is the _____ stage,
which occurs between _____. During this stage, infants acquire the mental ability to
understand that when objects are moved out of sight they still continue to exist. He termed this
ability _____ .

Piaget termed the second stage of cognitive development the _____ stage, a period of development experienced between the ages of _____. During this period, children generally think without using reason or logic. They find it difficult to understand that others may perceive the world differently from themselves. Piaget described this response as _____.

However, by the third stage of their cognitive development, the _____ stage (when they are aged _____), most children develop an understanding of the world that uses reasoning or logic. For example, they develop an understanding that mass, weight and volume remain unchanged when the shapes of objects change. Piaget termed the ability to think logically about concrete objects and situations _____ . According to Piaget, children this age also develop _____, which allows them to reverse their thoughts (or mental operations) about objects and situations.

When children reach the final stage of cognitive development, the _____ stage (which occurs from _____), Piaget suggested that children develop an ability to think in _____ . Now they can move away from thinking based on concrete objects and specific examples. They are able to consider future possibilities and incorporate _____ into their thinking patterns.

2.3.5 Erikson's theory of psychosocial development

Key science skills
Analyse, evaluate and communicate scientific ideas
- analyse and explain how models and theories are used to organise and understand observed phenomena and concepts related to psychology, identifying limitations of selected models/ theories

Develop

PART A

Draw lines between the matching stages of psychosocial development shown in the two columns below.

1	Trust	**A**	Guilt
2	Autonomy	**B**	Despair
3	Integrity	**C**	Shame and doubt
4	Industry	**D**	Role confusion
5	Intimacy	**E**	Stagnation
6	Generativity	**F**	Isolation
7	Identity	**G**	Mistrust
8	Initiative	**H**	Inferiority

PART B

Complete the flowchart in Figure 2.7 by filling in the major developmental tasks and life crises faced at each stage. Define the psychosocial dilemma in the space provided.

Figure 2.7 Major developmental tasks and life crises

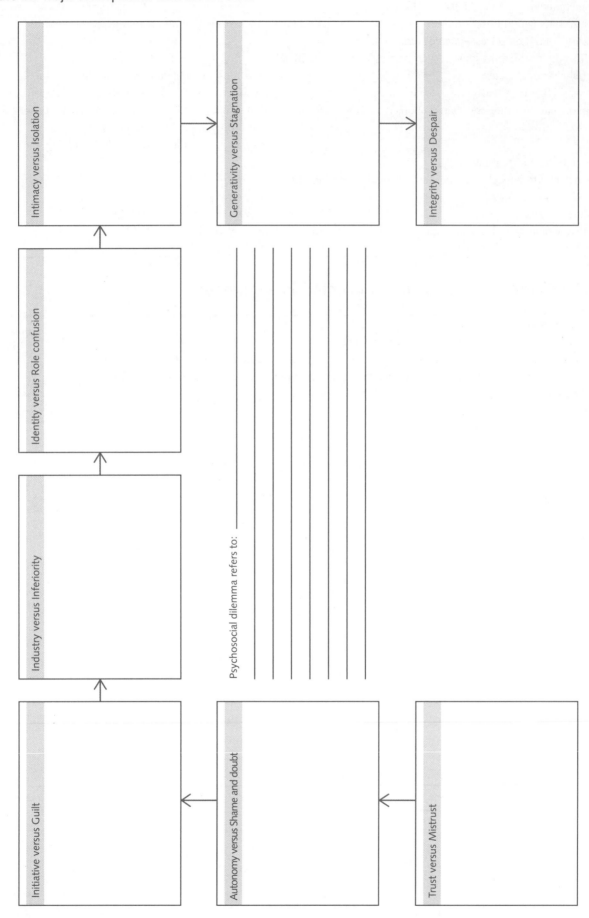

2.4 Sensitive and critical periods in psychological development

2.4.1 Comparing and contrasting critical and sensitive periods

Develop

1 Define 'critical period'.

_____

2 Define 'sensitive period'.

_____

3 Complete the following diagram in Figure 2.8 by comparing the similarities and differences between critical and sensitive periods. Write the differences in the circles under the appropriate headings, and the similarities in the section where they overlap.

> Similarities belong in the overlap between the two circles.

Figure 2.8 A Venn diagram of similarities and differences between critical and sensitive periods.

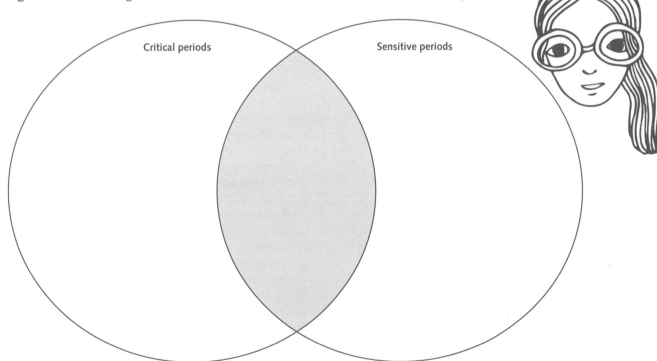

4 Give an example of a critical period and a sensitive period.

_____

_____

_____

2.4.2 Analysis of research

Key science skills

Develop aims and questions, formulate hypotheses and make predictions
- identify independent, dependent, and controlled variables in controlled experiments

Comply with safety and ethical guidelines
- demonstrate ethical conduct and apply ethical guidelines when undertaking and reporting investigations

Develop

Read the article below and answer the questions that follow.

MDMA may reawaken 'critical period' in brain to help treat PTSD

Johns Hopkins Medicine, *ScienceDaily*
April 4, 2019

Johns Hopkins neuroscientists have found that the drug MDMA reopens a kind of window, called a 'critical period', when the brain is sensitive to learning the reward value of social behaviors in mice.

Critical periods were first described in the 1930s in snow geese. About 24 hours after a gosling hatches, if the mother goose is nowhere to be found, the hatchling will bond with an object, including any non-living ones. Yet, if mother goose disappears 48 hours after her gosling hatches, the critical period is over, and the hatchling won't bond to an object.

There is evidence for critical periods that smooth the way for development of language, touch and vision.

For the current study, neuroscientist Gül Dölen says, 'We wanted to know if there was a critical period for learning social reward behaviors, and if so could we reopen it using MDMA, since this drug is well-known to have prosocial effects.'

Dölen and her team studied groups of mice in enclosures with different bedding. They put several mice together in one enclosure with one type of bedding for 24 hours and, in the next 24 hours, put the same mice by themselves in another enclosure with a different type of bedding. The mice began to associate certain types of bedding with isolation or companionship. Then, they let the mice wander between enclosures with the two types of bedding and tracked how long the mice spent in each enclosure. The more time the mice spent in the bedding linked to their companions indicated more social reward learning.

'It's why people gather around the water cooler,' says Dölen, assistant professor of neuroscience at the Johns Hopkins University School of Medicine. People are conditioned to know that the water cooler is an optimal place to chitchat with companions.

In their experiments, Dölen and her colleagues found that the critical period for social reward learning in mice is around puberty and wanes once they become mature adults. To determine if they could reopen the critical period, the scientists gave MDMA to mature mice, waited 48 hours for the drug to be washed out of their system, and observed how the mice explored their enclosure and behaved with other mice in the enclosure.

Following the treatment with MDMA, most of the animals responded to social interactions the same way as juveniles, by forming a positive association between social interactions and the bedding. This effect lasted for at least two weeks after the MDMA treatment, and it was not observed in mice given saline injections.

'This suggests that we've reopened a critical period in mice, giving them the ability to learn social reward behaviors at a time when they are less inclined to engage in these behaviors,' says Dölen.

Dölen and her postdoctoral student and first author of the current study, Romain Nardou, also observed that MDMA works to reopen the critical period only if the drug is given to mice when they are with other mice, not if it is given to mice while they are alone. This suggests that reopening the critical period using MDMA may depend on whether the animals are in a social setting, say the scientists.

The mice maintained their ability to learn the rewards of social behavior for up to two weeks from the time they were given MDMA. During this time, Dölen and her colleagues also found that the brains of the mice had corresponding responses to oxytocin, known as the 'love hormone', which is made in the hypothalamus and acts in the brain as a signal between neurons that encode information about social rewards. They found these responses by looking more closely at synapses, the spaces between brain cells called neurons. Their experiments showed that, in mature mice given MDMA, oxytocin triggers signaling in the synapses that encodes learning and memory, which does not typically happen in mature mice.

Dölen says that opening the critical window for social reward behavior may also have implications for treating psychiatric conditions.

MDMA has been designated by the U.S. Food and Drug Administration as a 'breakthrough therapy' for post-traumatic stress disorder (PTSD), meaning that the agency will fast-track the development and review of clinical trials to test it. However, the researchers caution that MDMA may not work for every psychiatric condition linked to social behaviours.

'As we develop new therapies or determine when to give these therapies, it's critical to know the biological mechanism on which they act,' says Dölen.

Source: Johns Hopkins Medicine. (2019, April 4). Psychadelic drug MDMA may reawaken 'critical period' in brain to help treat PTSD. *ScienceDaily*.

Questions

1 Identify the independent and dependent variables.

   a Independent variable

   _____

   b Dependent variable

   _____

2 Identify the experimental and control groups.

   _____

3 Name the experimental research design.

   _____

4 What is the purpose of using this design for this particular experiment?

   _____

   _____

5 According to this article, MDMA has been designated by the U.S. Food and Drug Administration as a 'breakthrough therapy' for PTSD, meaning that the agency will fast-track the development and review of clinical trials to test it. What ethical considerations should researchers follow for human clinical trials?

   _____

   _____

   _____

Chapter 2 summary

Indicate whether each statement in Table 2.10 is true or false by placing a tick in the correct column and writing the correction if you have chosen false.

Table 2.10 Psychological development concepts and theories

Statement	True	False	Correction
1 A critical period extends over a longer period of time than a sensitive period.			
2 Development that does not occur during a critical period will not occur later in life.			
3 Children learn a second language more easily than most adults. Adults can still learn a second language, but it may take more effort, because infancy is the most sensitive time to develop this skill.			
4 Ainsworth suggested that there are four phases in the development of attachment.			
5 Bowlby was the Canadian psychologist responsible for devising the 'strange situation'.			
6 Nature is more important than nurture in development.			
7 Twin studies are used to determine the influence of hereditary and environmental factors on development.			
8 Emotional development refers to how an individual thinks and solves problems.			
9 According to Piaget's theory, symbolic thinking is developed before concrete thinking.			
10 Piaget studied emotional development.			
11 There were eight stages in Kolberg's psychosocial development theory.			
12 The stages of cognitive development are, in order, sensorimotor, preoperational, concrete operations and formal operations.			
13 Adoption studies are used to determine the influence of hereditary and environmental factors on development.			
14 Identity versus role confusion is the seventh stage in psychosocial development.			
15 Normally there are 46 chromosomes in each cell of the human body.			
16 Identical twins can have the same genotype but different phenotypes.			
17 Hormones are chemical substances that are released from glands into the blood stream so that any imbalance in the physical system can be overcome.			
18 'Nature' refers to the environmental or external conditions that affect an individual's development.			
19 After birth, genetics cease to be an important factor influencing normal development.			
20 Harlow found that contact comfort was not as important as feeding in infant–mother attachment.			

Defining and supporting psychological development

3

3.1 Typical and atypical behaviour

3.1.1 What is typical behaviour?

Imagine a continuum from extremely atypical behaviour at one end to typical at the other. Use the statements from Table 3.1 and place them somewhere along the continuum on Figure 3.1, then answer the questions that follow.

We've all got our own little quirks!

Table 3.1 Examples of behaviours

Treating a pet as a baby	Believing that spirits have entered your body
Biting your nails	Binge eating
Belching after a meal	Streaking
Arguing with your teacher	Wearing dead husband's bones as a necklace
Leaving food on your plate	Tattooing or body piercing
Bungy jumping	Smiling all day
Men scratching their own genitalia	Believing you have been abducted by aliens
Washing your hands 10 times a day	Laughing at a funeral
Being breastfed at eight years old	Not changing your socks after exercise

Figure 3.1 Behaviour on a continuum

Atypical   Typical

1   What criteria did you use to influence your decision? Next to each statement write one of the following:
   - my culture
   - my personal experience/background.

2   What are the limitations of using a continuum to categorise behaviour?

_____

_____

_____

3.1.2 Typical or atypical?

Key science skills
Analyse, evaluate and communicate scientific ideas
- discuss relevant psychological information, ideas, concepts, theories and models and the connections between them

Develop

PART A

This activity will aid your understanding of the different psychological criteria used to determine whether behaviour is typical or atypical.

Choose from the following criteria: social norms, maladaptive behaviour, cultural perspective, personal distress, statistical rarity. Place your answer in the right-hand column of Table 3.2.

Table 3.2 Criteria used to determine whether behaviour is typical or atypical

Statement	Criterion
I make judgements depending on whether an individual can cope with everyday tasks such as washing, feeding themselves or going to work.	
I make judgements on whether behaviour is typical depending on the situation or the context in which the behaviour occurs. I take into consideration the popular standards of behaviour in a particular situation.	

⊗

»

I analyse whether the behaviours an individual is demonstrating are helping them to grow, develop and establish healthy relationships.	
I take into consideration the standards set by a society or a culture as to what is typical or atypical.	
I define normal behaviours as those that do not deviate from the average by very much. I make judgements based on how the majority of individuals behave, think or feel.	.

PART B

Read the following statements and decide which criterion you would use to make judgements as to whether the behaviour was typical or atypical. More than one criterion can be used for each statement.

Choose from the following criteria: social norms, maladaptive behaviour, cultural perspective, personal distress and statistical rarity. Place your answer in the right-hand column of Table 3.3.

Table 3.3 Applying criteria to behaviours

Statement	Criteria
When my grandmother was growing up it was not normal for women to wear jeans. However, it is seen as normal behaviour today.	
I lost my job because I couldn't get out of bed in the morning. I have not washed any of my clothes for more than two months.	
When I travelled overseas last year and visited some temples, I had to wear a long skirt and cover my shoulders. When I go to church in Australia I do not have to cover up.	
Lucy hears conversations in her head that nobody else can hear.	
Shandi has an IQ of 152.	
Joe feels 'down' almost every day. It upsets him that he doesn't feel like doing anything productive.	
I wear my pyjamas to bed, not to school.	
At a funeral last year, a man burst out laughing hysterically when everyone else was crying.	
Before the late 1800s, women in Australia could not vote in elections.	
Steven was diagnosed as suffering from depression.	
When my brother did his VCE he received an ATAR score of 88. This was well above average.	
My uncle Colin barks like a dog when there is a full moon, and he likes to sleep outside with the dogs.	
Ngarra has started a new job. His manager lets his work mates know that he finds direct eye contact during conversations disrespectful.	

3.1.3 Using statistical rarity to determine what is atypical

Key science skills

Generate, collate and record data
- systematically generate and record primary data, and collate secondary data, appropriate to the investigation

Analyse and evaluate data and investigation methods
- process quantitative data using appropriate mathematical relationships and units, including calculations of percentages, percentage change and measures of central tendencies (mean, median, mode), and demonstrate an understanding of standard deviation as a measure of variability

Develop

»

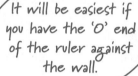

Analyse, evaluate and communicate scientific ideas
- use appropriate psychological terminology, representations and conventions, including standard abbreviations, graphing conventions and units of measurement
- analyse and explain how models and theories are used to organise and understand observed phenomena and concepts related to psychology, identifying limitations of selected models/theories

Develop

For this activity, choose a physical feature, such as foot length, to determine whether there are statistical rarities within your cohort.

MATERIALS

- ruler

INSTRUCTIONS

1 Determine your population of research interest.
2 Obtain a sample of participants.
3 Request participants remove their shoes, then place the heel of their foot against a wall.
4 Using a ruler, measure the length of the foot to the nearest centimetre (Figure 3.2)
5 Record the measurements on the data sheet (Table 3.4).

It will be easiest if you have the 'O' end of the ruler against the wall.

Figure 3.2 Method for measuring foot length

Table 3.4 Data sheet for measuring foot length

Participant	Male/female/other	Foot length (cm)
1		
2		
3		
4		
5		
6		
7		
8		
9		

»

10		
11		
12		
13		
14		
15		
16		
17		
18		
19		
20		

1　Complete a frequency table of the data.

2　Determine the following measures of central tendency:

- mean

- mode

- median

3　Analyse the data to determine outliers.

4　At what measure would you consider foot length atypical for your population of research interest?

5　What are the limitations of using a statistical approach to determine what is typical?

3.1.4 Adaptive or maladaptive?

Key science skills
Analyse, evaluate and communicate scientific ideas
- discuss relevant psychological information, ideas, concepts, theories and models and the connections between them

Develop

Transfer the statements from Table 3.6 to Table 3.5, categorising them as either adaptive or maladaptive.

Table 3.5 Applying the concept of adaptive and maladaptive

Adaptive		
Everyday performance of tasks that is required for a person to fulfill typical roles in society, including maintaining independence and meeting cultural expectations regarding personal and social responsibility (APA dictionary, n.d.).		

Emotions	Behaviours	Cognitions

Maladaptive		
Behaviours that interfere with optimal functioning in various domains, such as successful interaction with the environment and effectual coping with the challenges and stresses of daily life (APA dictionary, n.d.).		

Emotions	Behaviours	Cognitions

Table 3.6 Examples of behaviours

They understand emotions and recognise them in others.	They like to bully others.	They cheat on tests.
They drink alcohol every day.	If they have a problem, they talk to friends or family.	They self-harm.
They have repetitive thoughts and feelings about an event.	They express themselves through art/music/sport.	They don't leave their house.
They attend school.	They know how to seek help.	They withdraw from others when they have problems.
They feel emotions and can regulate them.	They always do things to seek attention.	They shower regularly.
They have an addiction.	They find joy in lots of things.	They laugh at bad news.
They exercise regularly.	They think about positive actions that help deal with a negative event.	They always talk over others.
They have unexpected and regular emotional outbursts.	They have acceptance over those things they can't control.	They like to risk-take.

3.2 Normality and neurotypicality

Key knowledge
- the concepts of normality and neurotypicality, including consideration of emotions, behaviours and cognitions that may be viewed as adaptive or maladaptive for an individual

3.2.1 Measuring emotions

Key science skills
Construct evidence-based arguments and draw conclusions
- discuss the implications of research findings and proposals, including appropriateness and application of data to different cultural groups and cultural biases in data and conclusions

Analyse, evaluate and communicate scientific ideas
- use appropriate psychological terminology, representations and conventions, including standard abbreviations, graphing conventions and units of measurement
- analyse and explain how models and theories are used to organise and understand observed phenomena and concepts related to psychology, identifying limitations of selected models/theories

Develop

Measuring emotions to make a judgement about what is considered typical or atypical is challenging. This is a simple activity that measures emotional reactivity.

1 Using the scale below make a judgement about your level of excitement.
- 0 = 'not at all'
- 50 = 'moderately'
- 100 = 'extremely'.

Next, imagine you haven't eaten all day and look at the image in Figure 3.3

Figure 3.3 An example of an image presented to participants to measure emotional reactivity

Shutterstock.com/88studio

Yummm!

2 What is your level of excitement now?

3 Calculate your reactivity by subtracting the first rating from the second.

My reactivity rating is _____.

4 What can your result tell you about your reaction to chocolate cake:

- if it was positive?

- if it was negative?

- if it was neutral?

5 How could this measurement tool be applied?

6 If you were to design an investigation using this tool, what considerations would you have to include to be culturally responsive and therefore be able to generalise your findings?

3.2.2 Personal stories

Key science skills

Analyse, evaluate and communicate scientific ideas
- discuss relevant psychological information, ideas, concepts, theories and models and the connections between them

Develop

MATERIALS

- highlighters

INSTRUCTIONS

Read the two personal stories and, using different colours, highlight examples of emotions, behaviours and cognitions that could be considered maladaptive for each individual.

Jane's story

I was very young when I experienced my first break from reality. I remember hearing voices and seeing shadows everywhere I went. Creatures of my mind. As a child, I was confused and scared of the hallucinations I was experiencing. I didn't understand why I was hearing and seeing the things I did. They would tell me that the world would benefit if I was no longer around or that I should harm someone just to protect myself. By the time I started the fifth grade, I experienced my first complete psychotic break.

One day at school, I became overwhelmed by the visions of shadow-like figures beginning to surround me. I felt so conflicted on what to do, it felt like all eyes were on me and everyone was out to get me and that I must protect myself. I ran out the classroom and hid inside of the girl's bathroom, locking myself in one of the stalls.

My teacher called the school guidance counselor and school police officer to calmly get me out of the stall. I remember screaming 'No, they're here! They are going to kill me.' They were obviously puzzled by who I meant was going to hurt me. I told them from behind the stall that the shadows and the man (the name of the voices I heard) was telling me to hurt others and myself. It took the police officer telling me that no one was coming to harm me and that I am much safer with him than alone by myself. I opened the bathroom stall and ran into the officer's arms and began to break down crying. I didn't know until after I came out the bathroom that there was a EMS team waiting for me with a stretcher. I didn't know that the hour I was in the bathroom with the guidance counselor and officer, that another counselor called my parents and they agreed to turn me into state care as they knew I was experiencing these symptoms, but had no clue where to take me or what to do.

I was transported to a nearby hospital where I met my parents. We together spoke to crisis intervention about the symptoms I had been experiencing and the next steps to take as a family. My parents talked it over with the interventionist and everyone agreed that I needed to stay inside of a hospital environment until I was better.

After a week of being in the psych hospital, I began to improve. My anti-psychotics were increased but I still was experiencing hallucinations and paranoid thoughts. The first time I heard my new diagnosis I was in family therapy at the hospital. The psychiatrist diagnosed me with early onset schizophrenia at the age of 11. I didn't understand that word. Schizophrenia was the thing that had been controlling my thoughts and haunting me since I was a young child.

It has been an on and off battle since being released from the hospital. The first few years were tough. I isolated myself from everyone else. I felt like an outcast. No one I knew was going through what I went through. As I stressed out about my social life, school and after school activities, I began to neglect taking my medication. Around this time, I was in middle school. Puberty was hard enough but being a preteen with a severe mental health diagnosis made life even harder for me to deal with.

I was 13 when I first attempted to take my own life. As I look on it now, I am happy I survived. But, it landed me another month back in the hospital. The doctors told me how important it was to take my medication. I took that advice to heart. I no longer wanted to be the victim of my diagnosis. I wanted to survive. It took a while, but I began to start taking my medication. Mainly, because I did not want to relapse again. I wanted to fight this.

With the help of my parents, therapist and school counseling staff, I am able to live with schizophrenia instead of letting it control my life. I began to interact more with my peers, I no longer felt alienated and I no longer let the hallucinations take charge of my life. I began to make more friends, my grades increased and I wasn't afraid of my own mind anymore. After graduating high school, I went into college to study not only Visual Arts but also Early Childhood Development. I feel like – regardless of a few setbacks – I am a recovery story.

If I would have to speak to myself at age 11 I would say, you are a strong young woman. Don't let fear consume your bright young mind. Get the help. There is nothing to be afraid of. Adults and professionals will help you through your hardest struggles. Don't isolate yourself. You are not alone.

'Living With Schizophrenia', NAMI, https://www.nami.org/Personal-Stories/Living-With-Schizophrenia.

Billy's story

One source of anxiety for me is knowing that people don't say what they mean. They skirt around the truth because they expect people to translate. This convention works well for some, because hearing the truth would be too painful; inferring the truth is gentler. The problem for me is I never know whether I've translated correctly. There's a constant uncertainty that makes it hard to trust people, even my closest friends and family.

In addition to my challenge with translating other people's words, I have to constantly consider the fact that people will inevitably translate my words even though I often mean things literally. It's as if people expect me to exaggerate because that's just part of modern-day human communication. I don't like to exaggerate one way or the other because I don't want my words to get translated. Only in my late twenties did I begin to understand these unwritten 'rules'. I'm learning how to soften my words before saying them, because they'll get translated or interpreted more harshly than I intend them. The misunderstandings

I've endured have left me with a sense that I need to walk on eggshells.

Another source of anxiety for me is my highly empathetic nature. I often feel as though the veils are thin between me and other individuals, and the general environment. Boundaries are weak; I get easily drawn into other people's needs and struggles. I feel their anxiety for them.

Poor boundaries and communication challenges have led to decreased confidence in my ability to maintain healthy relationships. Conflict and tension in my close relationships are stressful because they make me feel like my relationships aren't stable. The anxiety has a snowball effect. Interdependence is a need that everyone has, but especially people like me. Due to my sensory sensitivities, I am rarely able to drive, ride public transportation or go shopping. Furthermore, I am not able to work full time. I rely on others for housing, food sharing, transportation and assistance navigating the bureaucracies of modern-day society.

'My Autistic Anxiety Snowball' by Billy, https://www.aane.org/my-autistic-anxiety-snowball/. Adapted with permission from AANE

3.3 Variations in brain development

Key knowledge
- normal variations of brain development within society, as illustrated by neurodiversity

3.3.1 Who am I?

Key science skills
Analyse, evaluate and communicate scientific ideas
- discuss relevant psychological information, ideas, concepts, theories and models and the connections between them

Develop

Read the statements in Table 3.7 and determine which variation is being described. Write this in the column on the right. Your textbook will help with this.

Table 3.7 Normal variations of brain development

Description	Variation
I have difficulty with social interactions and changing from one activity to another. When completing an activity, I have a high attention to detail.	
I find it hard to sit still and regulate my emotions. I can be very inventive.	
I have tics and come out with random words and noises.	
I find it hard to control my muscles and move, this makes it hard to write and play sport. I can be good at problem-solving.	
Whilst I am good at seeing the big picture, I have difficulty with reading, spelling and writing.	
Please don't ask me to complete any maths calculations; numbers have no meaning for me.	
I have trouble getting organised, staying focused, making realistic plans and thinking before acting. I am fidgety, noisy and unable to adapt to changing situations.	

3.3.2 Embracing normal variations

Key science skills
Analyse, evaluate and communicate scientific ideas
- discuss relevant psychological information, ideas, concepts, theories and models and the connections between them
- analyse and evaluate psychological issues using relevant ethical concepts and principles, including the influence of social, economic, legal and political factors relevant to the selected issue

Develop

In Table 3.8, provide some suggestions on how our schools can adjust to a broader concept of what are normal variations of brain development. The first category has been completed as an example.

Schools can help to provide a flourishing evnrionment for everyone.

Table 3.8 Adjustments used to support neurodiversity

Category	Example	Adjustment
Sensory and motor	Light sensitivity Noise sensitivity Difficulty with fine motor skills such as handwriting	Provide short breaks during classes to help manage sensory sensitivities Offer noise cancelling headphones Provide light adjusting glasses Offer note taker for taking class notes
Cognitive	Easily distracted Miscomprehension due to literal interpretation Difficulty comprehending certain communication styles (verbal and gesture) Difficulties with new tasks or unplanned changes	
Behavioural	Poor organisational skills	
Social/interpersonal	Difficulties with group work Difficulties initiating or responding appropriately in communication with others	
Emotional	Feeling overwhelmed Anxiety and depression	

1   What are the issues for schools in providing these supports to cater for neurodiversity?

- economic

_____
_____
_____

- political

_____
_____
_____

- legal

_____

_____

_____

- social

_____

_____

_____

3.4   Supporting psychological development and mental wellbeing

Key knowledge
- the role of mental health workers, psychologists, psychiatrists and organisations in supporting psychological development and mental wellbeing as well as the diagnosis and management of atypical behaviour, including culturally responsive practices

3.4.1 Supporting mental wellbeing

Key science skills
Analyse, evaluate and communicate scientific ideas
- discuss relevant psychological information, ideas, concepts, theories and models and the connections between them

Develop

It's important to look after your mental wellbeing.

PART A

This activity will help to clarify your understanding of the links and distinctions between mental health, mental health problems and mental illness.

Complete the flowchart in Figure 3.4 by inserting a definition and an example of mental health, a mental health problem and mental illness.

_____

Figure 3.4 A comparison of mental health, a mental health problem and mental illness

Health:
A state of complete physical, mental and social wellbeing and not merely the absence of disease or infirmity (WHO, 1948).

Mental health

Mental health problem

Mental illness

PART B

Place a tick in the appropriate column in Table 3.9 to indicate which mental state each statement refers to.

Table 3.9 Applying the concepts of mental health and mental illness

Statement	Mental health	Mental health problem	Mental illness
Enjoying a sense of wellbeing, and being optimistic and emotionally resilient			
Experiencing a constant, unwanted thought that you cannot control and that negatively affects your daily functioning			
Wanting to stop gambling but feeling compelled to gamble every week			
Feeling very fearful of a situation because you have had a bad experience with it previously, and the thought that you cannot cope with its demands is with you constantly			
A significant decrease in your usual appetite and sleep patterns for a few weeks due to the stress of an upcoming exam			
Feeling stressed from the pressures of study but going to school every day and maintaining normal routines			
Withdrawing from your usual social activities and feeling sad for a few months after the death of your friend			
Feeling constantly anxious and apprehensive in a range of situations but not knowing why			
Having the ability to perform usual daily activities, but being unable to do so because of a persistent negative emotional state			
Occasionally experiencing mild compulsions and obsessive thoughts but being able to carry out normal daily routines			

PART C

1 Compare and contrast the roles of a psychologist and a psychiatrist by placing the descriptions provided in Table 3.10 into the appropriate section of Figure 3.5.

Figure 3.5 Comparison between roles of a psychiatrist and a psychologist

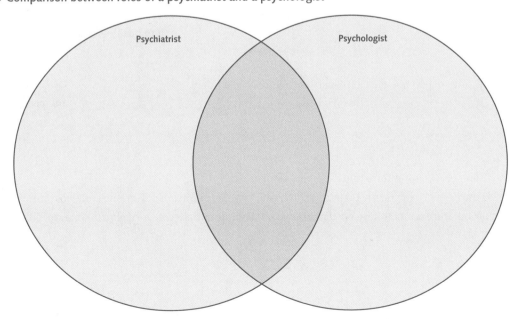

Psychiatrist   Psychologist

Table 3.10 Roles of a psychologist and psychiatrist

Prescribe medications	Specialise in mental health	Tertiary qualifications
Psychotherapy	Medical doctor	Psychological testing and evaluation
Mental and physical treatments		

2 Who would be the first person someone would see to obtain a referral to a psychologist or psychiatrist?

_____

3.4.2 Diagnosis of atypical behaviour

Key science skills

Analyse, evaluate and communicate scientific ideas

- discuss relevant psychological information, ideas, concepts, theories and models and the connections between them
- analyse and explain how models and theories are used to organise and understand observed phenomena and concepts related to psychology, identifying limitations of selected models/theories

Develop

PART A

This activity will improve your knowledge of how mental disorders are classified.

Use the terms in Table 3.11 to fill the gaps in the following paragraphs.

Table 3.11 Terms related to the classification of mental disorders

manual	social agencies	statistical	psychologists
consistent	diagnostic	categories	common
therapists	anxiety disorders	researchers	psychotic disorders
treatments	world		

The DSM-5 is the _____ and _____
_____ of Mental Disorders, published by the American Psychiatric Association. It groups
psychological problems into _____ based on similar symptoms in order to help diagnose and
treat mental disorders.

   The DSM-5 provides a _____ language for _____ ,
_____ , _____ ,
_____ and health workers, and is therefore used
all over the _____ . This allows for _____ diagnoses
and _____ .

   The major categories of psychological disorders are addiction disorders, _____ ,
_____ , mood disorders, _____ and personality
disorders.

PART B

1 Table 3.12 contains general symptoms and examples of mental disorders.
2 Match them to the correct category in Table 3.13. Each category has one set of symptoms and two examples.
3 Once the symptoms and examples have been matched to their appropriate category, glue them into Table 3.13.

Table 3.12 Symptoms and examples of mental disorders

Schizophrenia	Obsessive-compulsive disorder (OCD)	Poor functioning; inability to stop using drugs or engage in activity; emotional and relationship problems; experiencing withdrawal
Depressive disorder	Deeply ingrained, unhealthy personality patterns	Panic disorder
Delusional disorder	Disturbances in affect (emotion): mania, agitation, euphoria, hyperactivity or depression, sadness, hopelessness	Substance dependence: alcohol, barbiturates, opiates, cocaine, amphetamines, marijuana, nicotine
Feelings of fear, apprehension, anxiety; anxiety-based distortions of behaviour	Bipolar disorder	Anti-social personality disorder
Gambling dependence	Borderline personality disorder	Hallucinations, delusions, social withdrawal, retreat from reality, inability to control thoughts and actions

Table 3.13 Mental disorders, by category

Category	General symptoms	Examples of disorders
Addiction disorders		
Schizophrenia-spectrum disorders and other psychotic disorders		
Mood disorders		
Anxiety disorders		
Personality disorders		

3.4.3 Labelling and stigma

Key science skills

Analyse, evaluate and communicate scientific ideas
- discuss relevant psychological information, ideas, concepts, theories and models and the connections between them
- analyse and evaluate psychological issues using relevant ethical concepts and principles, including the influence of social, economic, legal and political factors relevant to the selected issue

Develop

In this activity you will further explore the concepts of normality and mental illness, and the classifications used to diagnose mental disorders. You will also explore the concept of labelling.

In 1973, David Rosenhan and his colleagues set out to investigate whether staff at mental institutions could tell the difference between the 'sane' and the 'insane'. Read about the study and answer the questions that follow.

Rosenhan's study of 'sanity' and labelling

Rosenhan and colleagues had eight 'sane' people gain entrance to 12 different hospitals for mentally ill people. None of the eight individuals (pseudo-patients) had ever experienced any serious symptoms of mental illness. Of the eight pseudo-patients, three were female and five were male. The youngest was a psychology graduate in his 20s; the others were older. Among the pseudo-patients were three psychologists, a paediatrician, a psychiatrist, a painter and a housewife.

The pseudo-patients each gained entry to a hospital stating that they were hearing voices. They told admissions staff at the hospitals that they heard voices of the same gender as themselves and that the voices were often unclear but at times could be heard saying words such as 'empty', 'hollow' and 'thud'. All eight pseudo-patients were admitted to the various hospitals when they presented with this one symptom. All but one was diagnosed as suffering from schizophrenia.

However, beyond their initial declaration that they were hearing voices, they behaved as they would normally, and continued to do so when they were admitted. They used fake names and some of them used fake employment details (the psychiatrist and psychologists did not give these as their professions); however, all other details about personal history and traits were presented as they actually were. None of the histories were seriously pathological in any way.

While in the hospitals, the pseudo-patients followed instructions from attendants, were administered medications (but did not swallow them) and spent some time writing down their observations. They initially wrote secretly, but after some time it became apparent that nobody cared whether they were constantly writing, so they started to do it in public. Presumably their writing behaviour was considered part of their mental disorder.

The length of stay in the hospitals varied from 7 days to 52 days, with the average being 19 days. The pseudo-patients entering the hospitals were not told when they would be released – each was told that they would be released when they were well. In other words, they had to convince staff members that they were 'sane'.

Ironically, many of the true patients in the hospitals did not believe the pseudo-patients had any kind of mental disorder – 35 out of 118 true patients voiced their suspicions, stating things such as, 'You are not crazy', 'You're a journalist, or a professor' (referring to the note-taking), or 'You're checking up on the hospital'.

Each of the pseudo-patients managed to gain release. They were all released on the basis that they were 'in remission' from their mental disorder.

Rosenhan and colleagues concluded that staff who were trained to detect mental illness and abnormal behaviour could not recognise normal behaviour once a label had been assigned to an individual stating that they suffered from a mental disorder. It appeared that once the initial impression had been formed, the psychiatric label had a life and influence of its own, so that once labelled as schizophrenic, there was an expectation that an individual would continue to be schizophrenic even if they did not show any symptoms.

When Rosenhan published the results, staff at a research and teaching hospital doubted that such errors could happen in a hospital. Rosenhan consequently conducted a follow-up study at the hospital where the doubting staff worked. Rosenhan informed the hospital that a number of pseudo-patients would attempt to admit themselves to the hospital over a 3-month period, but did not tell them exactly when. Over this time, each staff member was asked to rate all patients being admitted to the hospital on a 10-point scale,

9780170465045

showing whether they thought the patient was legitimate, or a pseudo-patient faking their symptoms.

Of the 193 patients who were admitted for treatment during that time, 41 of them were considered to be suspected pseudo-patients by at least one member of staff, while 19 patients were suspected of being 'sane' by at least one psychiatrist and one other staff member. However, unbeknownst to the hospital staff, no pseudo-patients were actually sent to the hospital by Rosenhan and colleagues during this time!

These findings have been criticised by others, but the study does raise important questions regarding the treatment of individuals suffering from mental disorders. A person cannot be a disorder; they can only suffer from one. Labelling can be detrimental to people who suffer from a mental disorder, though with improved education, the effects of labelling and the stigma attached to mental disorders may be reduced in the future.

Source: adapted from Rosenhan, DL. (1973). On being sane in insane places. *Science*, 179, 250–258.

1 Identify the aim of the initial study.

2 Describe the participants in the initial study.

3 Describe the experimental procedure used in this study.

4 What were the results of the initial study?

5 Rosenhan and his colleagues used some deception in these studies. Why? Was this ethical?

6 What was the purpose of Rosenhan and his colleagues conducting a follow-up experiment?

_____
_____
_____
_____
_____
_____

7 How do these studies suggest that stigma can be reduced in the future?

_____
_____
_____
_____
_____
_____

3.4.4 Managing atypical behaviour

Key science skills

Analyse, evaluate and communicate scientific ideas
- discuss relevant psychological information, ideas, concepts, theories and models and the connections between them
- acknowledge sources of information and assistance, and use standard scientific referencing conventions

Develop

For this activity, choose a specific mental disorder to research from the list in Table 3.14. If you are interested in a disorder that is not listed, you may be able to negotiate with your teacher to research that disorder instead.

If it's not on the list, check with your teacher first.

Table 3.14 Categories and examples of mental disorders

Category	Examples of disorder
Psychotic disorders	Schizophrenia
Mood disorders	Bipolar disorder, depressive disorder (major depression)
Anxiety disorders	Phobias, obsessive-compulsive disorder, post-traumatic stress disorder
Addictive disorders	Gambling, substance-related disorder
Personality disorders	Paranoid personality disorder, anti-social personality disorder

Use the internet, textbooks and any other available resources to complete the following research task questions. Here is a list of relevant websites you might like to visit:

- www.sane.org
- www.mentalhealth.com
- www.mental-health-matters.com
- www.planetpsych.com
- www.nimh.nih.gov
- www.beyondblue.com.au
- www.headspace.org.au
- www.embracementalhealth.org.au

1   Write down the general facts about the disorder, including:

- its definition

_____

_____

- the category it belongs to

_____

- the prevalence and demographics of the disorder (for example, age and gender statistics).

_____

_____

2   Describe the symptoms and behaviours that are present in someone with the disorder.

_____

_____

_____

_____

3   List the possible causes for the disorder, including:

- biological and environmental causes

_____

_____

_____

- risk factors for the disorder.

_____

_____

_____

4   Describe the treatment(s) that may be effective and/or that is usually administered for the disorder under the following categories:

- psychological

_____

_____

_____

- behavioural

_____

_____

_____

- social.

_____

_____

_____

5 Record your sources of information in a bibliography or reference list.

_____

_____

_____

_____

_____

_____

Chapter 3 summary

Read the following case study and answer the questions that follow.

Karen was a 37-year-old Jamaican-American female and a single mother of three teenage children, who had recently lost her job. Karen presented to our cognitive-behavioral treatment (CBT) clinic with a primary diagnosis of panic disorder with significant agoraphobia, and additional diagnoses of obsessive-compulsive disorder and generalized anxiety disorder. She had also had a past history of major depression and post-traumatic stress disorder from chronic and multiple traumatic experiences.

The primary aim in the first several sessions was to fully explore the role of cultural beliefs in the development and maintenance of Karen's symptoms. Within the first session itself, Karen expressed her strong religious belief and heavy involvement in church. Related to this, it was clear that Karen received many negative messages from her children, mother, and church friends about both her experience of panic-symptoms, and her decision to receive 'outside' (i.e. outside of the Caribbean-American community) psychological help. This did not deter Karen from seeking treatment, but through therapist exploration, Karen admitted that this certainly fuelled her own negative beliefs about being different from everyone around her and made her feel discouraged about ever becoming better. She also felt depressed about not being able to 'kick these symptoms' on her own simply through prayer and faith as others suggested and felt like a failure about this perceived deficiency.

With these larger cultural themes in mind, the therapy content started focusing on specific anxiety symptoms, and explored how culture infused her psychological symptoms in more detail. For instance, Karen reported that her obsessive thoughts about being poisoned by others (which would result in avoidance of eating or drinking items given to her by others at their homes, or in other settings outside her own home) stemmed from a strong belief in black magic, and that others were trying to harm her out of jealousy and control by the devil.

The strong belief in black magic and having 'an evil eye' cast upon someone was recognized as a widely accepted, culturally-congruent belief in individuals from the Caribbean, and the therapist therefore did not question the validity of this belief.

Similarly, given the expressed importance of religion in Karen's life view, the therapist explicitly incorporated Karen's spiritual strengths into the treatment very early on, particularly to target the stigma she experienced from others surrounding her decision to pursue psychological treatment. Specifically, the therapist proposed the idea that Karen's decision to seek formal help to address her anxiety symptoms was an example of her following her own internal spiritual compass in order to maximize her strengths and abilities to contribute to her community and family.

A recurrent theme throughout treatment was the importance of belief in God and service to her church. Consequently, this particular community/belief system was often referred to and integrated into the homework exercises and therapy discussions. Aside from its previously described use to increase her motivation to address the anxiety symptoms that were interfering with her meeting her full potential, the use of her community and church involvement were utilized. For instance, as she was nearing the end of treatment, she independently volunteered to work with the clinic administrators to disseminate information about available services (e.g. by going into local churches in her neighbourhood to share her positive experiences with therapy and anxiety reduction). This exercise was extremely empowering for Karen and met this need to contribute meaningfully to her community. There was also a focus on strengthening interpersonal relationships with her family members (e.g. going to get her nails done with her young adolescent daughter), because of her highly expressed cultural value of staying close-knit as an immigrant family. Thus, spending time with such individuals served both as an exposure exercise (to reduce agoraphobic avoidance) and to meet this valued interdependent cultural goal.

There appeared to be a plateau reached in several of her symptoms (particularly her chronic worry symptoms), and therefore it was decided that first-line CBT skills needed to be supplemented with other related techniques. Of note, the technique of mindful meditation was presented to Karen. However, to make this technique more palatable to this client, mindfulness activities that emphasized resilience and spiritual values-driven mindfulness practices were presented. Karen really enjoyed these exercises and reported a significant relief in the frequency and severity of her chronic worry through consistent use of this skill.

Termination of therapy was collaboratively decided upon, on the basis of significant reduction in anxiety symptoms. At the last session, Karen brought in a gift (a velvet rose) and a card with religious themes as a gesture of gratitude for the therapist. At this particular treatment setting, gifts are usually declined, but the therapist regarded Karen's gift choice as a noted effort to stay within the discussed boundaries and recognized the card as her desire to express herself from within her own belief system. Given this, the token was accepted by the therapist.

Source: Asnaani, A. and Hofmann, S.G. (2012), Collaboration in Multicultural Therapy: Establishing a Strong Therapeutic Alliance Across Cultural Lines. J. Clin. Psychol., 68: 187–197. https://doi.org/10.1002/jclp.21829

1   Was Karen seeing a psychiatrist or a psychologist? Justify your choice.

_____

_____

2   Where on the mental health continuum does Karen lie? Justify.

_____

_____

3   Which category or categories from the DSM-5 does Karen fit under?

_____

_____

4   What is stigma and how does this affect Karen?

_____

_____

_____

5   How has Karen's atypical behaviour been managed?

  • Psychologically?

_____

_____

  • Socially?

_____

_____

  • Culturally

_____

_____

6   Suggest two organisations that Karen can access to assist her and outline the services they provide.

_____

_____

_____

_____

The role of the brain in behaviour and mental processes

4

4.1 How the brain controls behaviour and mental processes

Key knowledge
- different approaches over time in understanding the role of the brain in behaviour and mental processes

4.1.1 History and concepts of brain research

Key science skills
Analyse, evaluate and communicate scientific ideas
- discuss relevant psychological information, ideas, concepts, theories and models and the connections between them

Develop

Use the terms and names in Table 4.1 to answer the following 'Who am I?' questions.

Table 4.1 Terms related to the history of behaviour and the brain

phrenology	René Descartes	philosophy	Franz Gall	Plato
lobotomy	vivisection	split-brain operations	neuroimaging techniques	Pierre Flourens
Aristotle	lesioning	Aeilius Galen	dualism	

Who am I?

1 I have many theories about human behaviour, but I don't have any empirical evidence to back them up. _____

2 Because I provide visual evidence of brain structure and function, neuroscientists can identify changes in a person's brain that otherwise could not be seen without surgical invasion. _____

3 I argued that the mind was located in the heart, which I suggested was the body's most important organ. _____

4 I suggested that because the brain is a physical substance it can be measured and observed and, like all living physical substances, it will eventually die. The mind, however, is a mental substance that, although it is located in the brain, is not physical. It cannot be measured, and it does not die when the physical body dies. _____

5 I am the first person associated with conducting public vivisections on animals to demonstrate that the mind is located in the brain and that the brain controls all behaviour. _____

6 I am a scientific or medical practice involving experimentation on live animals that involves cutting into or physically damaging them to demonstrate how internal structures operate. _____

7 My theory is that the mind and body are separate but parallel entities that interact through the brain's pineal gland. _____

8 I am the theory that the mind and body are separate entities that interact and influence each other. _____

9 According to me, feeling the contours of your skull will enable me to accurately assess your personality type and intelligence level. _____

10 I am the person mainly associated with introducing the theory of phrenology during the l9th century. _____

11 I am the French scientist who electrically stimulated or removed parts of animals' brains to demonstrate that phrenology's claims were false. _____

12 I am a form of brain surgery that involves severing nerves in the frontal lobe associated with emotional control to permanently alter a person's behaviour. _____

13 I am a modern surgical procedure involving lasering, surgical removal or vaporisation to remove or destroy brain tumours or change brain circuits. _____

14 I am used to interrupt the flow of information between the left and right hemispheres for people who suffer from extreme epileptic seizures. _____

4.1.2 Split-brain operations

Key science skills

Analyse, evaluate and communicate scientific ideas

- discuss relevant psychological information, ideas, concepts, theories and models and the connections between them

Develop

PART A

Label Figure 4.1 with the following terms in Table 4.2.

_____

Table 4.2 Terms related to split-brain operations

left visual fields	right primary visual cortex
optic chiasm	optic tract

Figure 4.1 Representation of split-brain operations

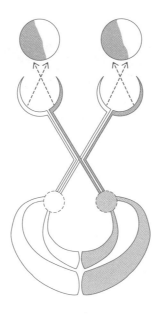

PART B

Use the images in Table 4.3 to answer the related questions.

Table 4.3 Analysis of the effects of split-brain operations

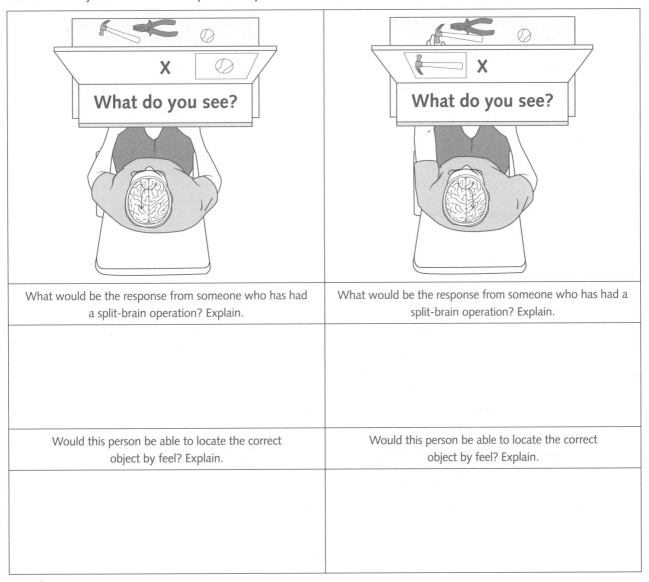

What would be the response from someone who has had a split-brain operation? Explain.	What would be the response from someone who has had a split-brain operation? Explain.
Would this person be able to locate the correct object by feel? Explain.	Would this person be able to locate the correct object by feel? Explain.

4.2   Brain structure and function

Key knowledge
- the roles of the hindbrain, midbrain and forebrain, including the cerebral cortex, in behaviour and mental processes

4.2.1 Hindbrain structures and functions

Key science skills
Analyse, evaluate and communicate scientific ideas
- discuss relevant psychological information, ideas, concepts, theories and models and the connections between them

Develop

1   Match the terms and definitions of hindbrain structures from Table 4.4.

2   Write the correct term with its definition in the appropriate labels on to the brain diagram (Figure 4.2) in the correct location.

Table 4.4 Terms for and definitions of hindbrain structures

Medulla	A group of nerves that connects the cerebral cortex (the outermost layer of the cerebral hemispheres) with the medulla. It is involved in arousal, sleep, daydreaming, waking, breathing and coordination of some muscle movements and motor tone.
Cerebellum	A hindbrain structure, located at the base of the brainstem, responsible for regulating internal bodily systems necessary for survival, such as heart rate and breathing.
Pons	A wrinkly structure attached to the rear of the brainstem that helps coordinate voluntary movement and balance.

Figure 4.2 Hindbrain structures and functions

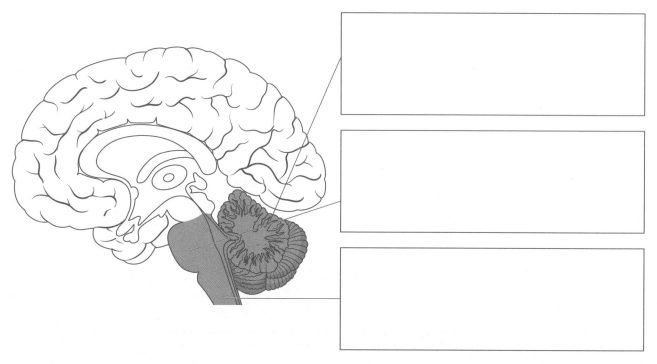

4.2.2 Forebrain structures and functions

Match the terms and definitions of forebrain structures from Table 4.5, then use the pairs to label the brain diagram (Figure 4.3).

Table 4.5 Terms for and definitions of forebrain structures

Thalamus	The two large cerebral hemispheres that cover the upper part of the brain.
Cerebrum	A peanut-sized structure in the forebrain responsible for a variety of functions, including regulation of body temperature, sex drive, appetite and the sleep–wake cycle. Communicates with the pituitary gland.
Hypothalamus	A structure that sits on top of the brainstem, through which all sensory information (except smell) passes; this brain structure redirects this information to the appropriate sensory area of the cerebral cortex for processing.

Figure 4.3 Forebrain structures and functions

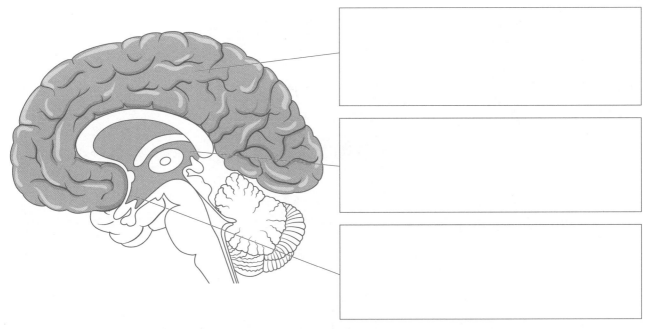

4.2.3 The brain communicates with the endocrine system

Label Figure 4.4 with the following terms from Table 4.6.

Table 4.6 Terms related to the endocrine system and its communication with the brain

hypothalamus	pituitary gland	adrenal glands
pancreas	thyroid	testes
ovaries	pineal gland	

Remember that:

- The hypothalamus stimulates the pituitary gland to release hormones that further stimulate the adrenal glands to release the hormone cortisol during times of stress.
- The pituitary gland stimulates the release of the hormones ghrelin and leptin, which are involved in the sensation of hunger.
- The pineal gland is stimulated to release melatonin, a hormone that induces sleep.

Figure 4.4 Major glands of the endocrine system

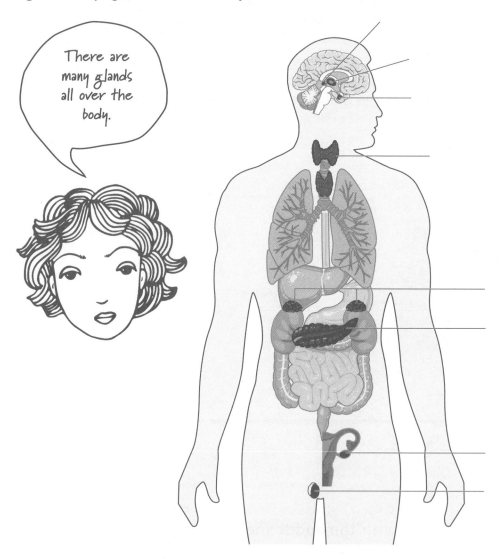

4.2.4 Structures connecting the hindbrain and forebrain via the midbrain

Develop

1 Match the terms for and definitions of midbrain structures that connect the hindbrain and forebrain from Table 4.7, and use these to label Figure 4.5. As a reference the thalamus (part of the forebrain) has been labelled for you.
2 Draw arrows in one colour to show the extensions of the reticular formation into the cerebrum.
3 Draw arrows in another colour to show the extensions of the basal ganglia into the cerebrum, where motor movement is coordinated.

Table 4.7 Terms for and definitions of midbrain structures

Reticular formation	A network of neurons, extending from the top of the spinal cord up to the thalamus, which modulates incoming sensory stimuli and redirects it to the cerebral cortex, thereby activating the cortex and influencing our state of physiological arousal and alertness.
Basal ganglia	Clusters of nerve cells deep within the cerebrum that transmit motor messages to and from the cerebral cortex. The basal ganglia are situated at the base of the forebrain and top of the midbrain. Basal ganglia are strongly interconnected with the cerebral cortex, thalamus and brainstem.

Figure 4.5 The reticular formation and basal ganglia connect the hindbrain to the forebrain

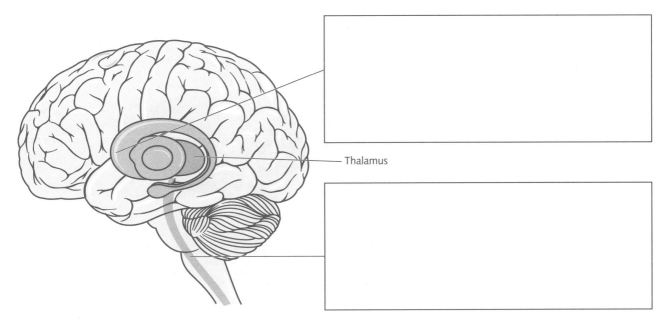

Thalamus

4.2.5 Forebrain, midbrain and hindbrain structures

Key science skills
Analyse, evaluate and communicate scientific ideas
• discuss relevant psychological information, ideas, concepts, theories and models and the connections between them

Develop

Assign the brain structures in Table 4.8 to Figure 4.6.

Table 4.8 Main structures of the forebrain, midbrain and hindbrain

pons	thalamus	reticular formation	cerebrum
cerebellum	medulla	basal ganglia	hypothalamus
forebrain	midbrain	hindbrain	

Figure 4.6 Main structures of the hindbrain, midbrain and forebrain

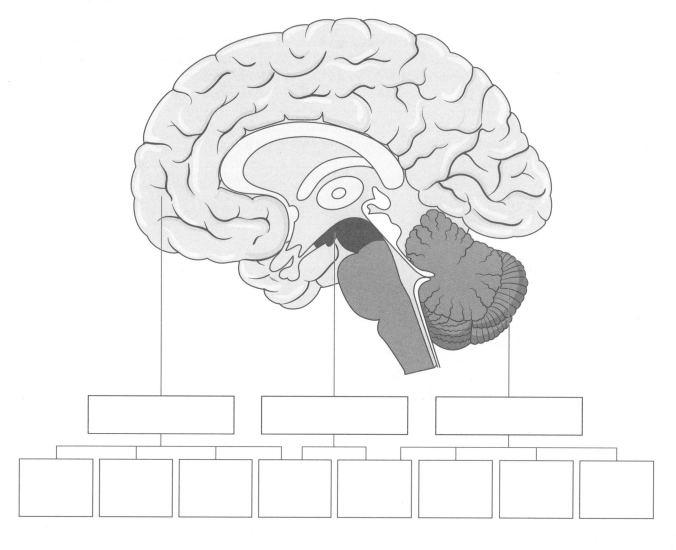

4.2.6 Functions of the hindbrain, midbrain and forebrain: summary

Key science skills
Apply, evaluate and communicate scientific ideas
- discuss relevant psychological information, ideas, concepts, theories and models and the connections between them

Develop

PART A

Use the terms in Table 4.9 to fill in the blanks in the following paragraphs. Terms may appear more than once.

Table 4.9 Forebrain, midbrain and hindbrain terms

alertness	thalamus	right	forebrain	balance
smell	corpus callosum	hypothalamus	medulla	cerebral cortex
cerebellum	endocrine system	brain stem	organs	cerebrum
hindbrain	basal ganglia	reticular formation	midbrain	left

Structurally, the human brain can be divided into three main regions: the hindbrain, _____
and forebrain. Each region has its own functions, and each is vital for everyday functioning and information processing.

The hindbrain is often referred to as the _____, because it is located at the base
of the brain near the back of the skull. It is an important part of the brain: it controls basic survival functions, as well as
coordinating voluntary muscle movements. The _____ is one hindbrain structure that
plays an essential role, because it regulates the function of internal _____
vital for life, controlling such things as heart rate. The _____ is
another _____ structure and is often referred to as the 'little brain'. Located at
the rear of the brain stem, it helps us to coordinate our voluntary movement and _____ .

The midbrain is the area of the brain between the hindbrain and the forebrain. It includes systems that help
to keep us alert, awake and vigilant. The network of neurons extending from the top of the spinal cord up through
the brainstem is known as the _____. This brain structure is responsible
for stimulating the cortex, arousing it to a state of _____ and activity.
The _____ sits below the forebrain and above the midbrain and is responsible primarily
for motor control, as well as other roles such as motor learning, executive functions and behaviours, and emotions.

The _____ is the largest and most highly developed part of the brain. One peanut-
sized structure, the _____, is located in this part of the brain. It plays a significant role in
controlling many aspects of our body, such as body temperature, our sleep–wake cycle, sex drive and our thirst and hunger.
This cluster of neurons works by communicating and connecting the nervous system
and _____ in order for our bodies to achieve a specific physiological state. Another brain
structure in this region is the _____, which acts as a relay system for all sensory messages
on their way to the cerebral cortex, except for _____. The largest brain structure in this
region, the _____, is divided into two hemispheres. The left and right hemispheres do not
actually touch; instead, they are connected by a bundle of nerve fibres, the _____ .
The outer layer of the cerebral hemispheres is the _____ and it enables our most
complex brain processes, such as emotions, motivations, sensations, perceptions, learning, memory and reasoning.
The _____ hemisphere is predominantly associated with verbal and analytical functions;
whereas the _____ hemisphere specialises in spatial tasks and visual skills, such as
recognising patterns and faces.

PART B

Read each scenario and determine which forebrain, midbrain or hindbrain structure is responsible for each function.

1 I am said to be the brain's internal 'body clock'. Without me your circadian rhythms may be dysfunctional.
What am I? _____

2 When visual information is received from the eyes, this sensory information passes through me, and I relay it to the appropriate area of the cerebral cortex (visual cortex). What am I? _____

3 I am responsible for regulating internal bodily systems necessary for survival, such as heart rate and breathing. What am I? _____

4 I enable verbal communication. Without me you would not be able to speak. What am I? _____

5 I am involved in arousal, sleep, daydreaming and waking. What am I? _____

6 I enable you to selectively focus your attention, the way you are now focusing on answering this question. What am I? _____

7 I direct the pituitary gland to release hormones, such as prolactin (associated with breastfeeding) and hormones that further stimulate the release of cortisol (associated with stress). What am I? _____

8 I enable smooth, coordinated movements, posture and balance; for example, allowing a gymnast to perform an acrobatic routine on a balance beam. What am I? _____

9 I enable information to be processed holistically; you are able to put together a puzzle because of me. What am I? _____

10 I help to fine tune motor movement. I am also involved in learning and the processing of emotion. What am I? _____

4.3 The cerebral hemispheres

Key knowledge
- the roles of the hindbrain, midbrain and forebrain, including the cerebral cortex, in behaviour and mental processes

4.3.1 Hemispheric specialisation

Key science skills
Apply, evaluate and communicate scientific ideas
- discuss relevant psychological information, ideas, concepts, theories and models and the connections between them

Develop

1 Write a definition of hemispheric specialisation in the box provided at the centre of Table 4.10

2 The bullet points provided indicate dominant cognitive and behavioural functions of the left and right hemispheres of the human brain. Write each bullet point in the correct part of Table 4.10.

- Controlling movement of voluntary muscles on left side of body
- Production of speech (Broca's area)
- Reading body language
- Processing information holistically
- Detecting and expressing emotion
- Registering sensory information from left side of body
- Logical reasoning
- Mathematical skill
- Processing information sequentially
- Spatial ability
- Pattern recognition
- Understanding of language (Wernicke's area)
- Registering sensory information from right side of body
- Musical ability
- Analytical thinking
- Controlling movement of voluntary muscles on right side of body
- Writing
- Facial recognition

Table 4.10 Terms related to specialisations of the left and right hemispheres

Left hemisphere	Right hemisphere
Sensory functions	Sensory functions
Motor functions	Motor functions
Cognitive functions	Cognitive functions
Perceptual functions	Perceptual functions

Hemispheric specialisation

4.3.2 The roles of the cerebral cortex

Key science skills

Analyse, evaluate and communicate scientific ideas
- discuss relevant psychological information, ideas, concepts, theories and models and the connections between them

Develop

1   Indicate whether each statement is true or false by placing a tick in the correct column in Table 4.11.
2   If the statement is false, underline the incorrect section and identify the reason why it is incorrect in the 'Correction' column (see Statement 1 for an example).

Table 4.11 Statements relating to the roles of the cerebral cortex

Statement	True	False	Correction
The cerebral cortex is responsible for higher cognitive and emotional functioning. It is composed of densely packed nerve cells (grey matter) that form the spongy, wrinkled outer layer of the brain that is <u>approximately 3–4 cm thick</u>.		✔	Cerebral cortex is 3–4 mm thick.
Damage to the rear strip of the frontal lobe could cause a person to have inappropriate emotional responses, difficulty concentrating and making decisions, and an inability to think logically.			
A function of the prefrontal cortex (frontal lobe) is emotional regulation.			
The primary auditory cortex, located in the parietal lobe, processes sounds.			
The motor cortex, located at the front of the parietal lobe, directs the body's voluntary muscles (skeletal muscles).			
Symbolic thinking (the ability to mentally represent people, objects and events using abstract concepts such as words, gestures and numbers) is a function of the occipital lobe.			
Broca's area (which in most people is located in the left frontal lobe and close to the strip of motor cortex) controls the muscles responsible for the production of fluent, articulate speech.			
On the motor cortex, the area dedicated to specific body parts represents the sensitivity of that body part. Therefore, because lips are more sensitive than elbows, the area relating to lips will be greater on the motor cortex than the area relating to elbows.			
The outer layer of the cerebral cortex is deeply folded and wrinkled. These convolutions mean that a very large surface area of the cortex can fit inside the human skull, allowing the space for many possible neural connections to be made.			
The primary visual cortex, located in the occipital lobes, processes visual information detected by the sensory receptors (photoreceptors) in our eyes.			

⟫

9780170465045

》

On the somatosensory cortex, the area dedicated to body areas depends on the number of muscles involved in moving the specific body part. Therefore, because more muscles are required to move your fingers than your knee, the area controlling your fingers occupies more space on the somatosensory cortex than the area for your knee.			
The primary auditory cortex of each temporal lobe receives and processes sounds sent from both ears.			
The area of cells in the left temporal lobe that is responsible for the understanding of language (written and spoken) is Wernicke's area.			
The arcuate fasciculus is a bundle of nerve fibres connecting the two language centres: Broca's area and Wernicke's area.			
Association areas of the frontal lobe receive information from other brain areas and structures, integrate it and determine responses to it.			
The basal ganglia are the bundles of nerve fibres connecting the left and right cerebral hemispheres.			
The primary visual cortex, located in the temporal lobe, registers visual stimuli sent from the retinas in your eyes.			
Psychologists believe that the frontal lobes are heavily involved in emotional behaviour and personality.			
The temporal lobes are heavily involved in memory and assist us to recognise faces and identify objects.			

4.3.3 Brain poster

Key science skills

Apply, evaluate and communicate scientific ideas

• discuss relevant psychological information, ideas, concepts, theories and models and the connections between them

Develop

MATERIALS

- a large sheet of poster cardboard (at least A3)
- scissors
- highlighters or coloured pencils
- glue
- pen

INSTRUCTIONS

1 Label the brain areas indicated in Figure 4.7.

2 Cut out Figure 4.7 and glue it onto the middle of your poster cardboard.

3 Cut out the descriptions of areas of the brain in Table 4.12 and arrange them around the figure of the brain on your poster, close to their corresponding brain area.

4 When you are satisfied with your layout, glue the information cells to the poster and draw arrows that connect the information to the appropriate part of the brain diagram.

5 Use highlighters or pencils to colour-code the lobes of the brain.

Figure 4.7 Areas of the brain

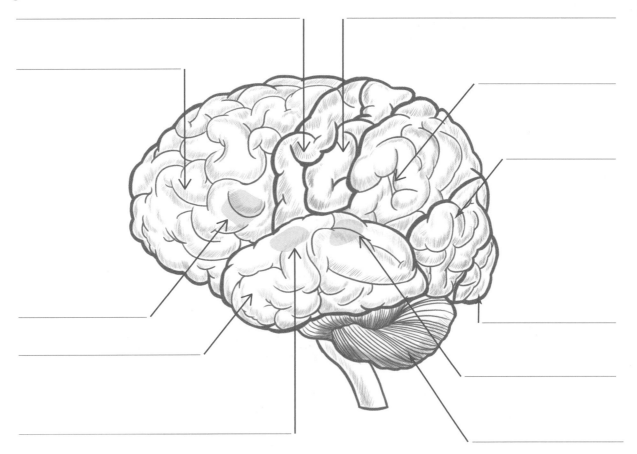

Table 4.12 Descriptions of areas of the brain

Frontal lobe	Parietal lobe
Association areas register information sent from other brain areas, integrate it and determine the response	Somatosensory cortex registers bodily sensations on the opposite side of the body (sent from muscles, joints, glands and organs). The amount of somatosensory cortex devoted to each body part corresponds to the sensitivity of the body part, not its size
Responsible for higher cognitive behaviour (e.g. decision-making, planning, problem-solving, analysing)	
Responsible for emotional behaviour and personality	Association areas register sensory information sent from the somatosensory cortex and other brain areas and coordinates it
Motor cortex directs the body's skeletal muscles (on the opposite side of the body) so it controls voluntary movement. The amount of motor cortex devoted to each body part corresponds to the importance of bodily areas, not their size	Monitors the body's position in space
	Responsible for spatial perception and locating objects in space

⟫

Wernicke's area (a language centre)	Temporal lobe
Responsible for understanding language (spoken and written) and giving meaning to sounds and words; therefore, involved in naming objects	Primary auditory cortex registers auditory information sent from both ears and determines pitch, rhythm and volume of sounds
Formulates grammatically correct, coherent and understandable speech	Association areas determine the nature of the sound (i.e. speech, music or general noise)
Locates appropriate words from memory to express meaning	Involved in memory processes (stores sound memories)
	Responsible for recognition (naming) of visual forms
Occipital lobe	**Broca's area (a language centre)**
Primary visual cortex registers visual information sent from the retinas of both eyes	Controls speech muscles, so controls production of articulate (clear and fluent) speech
Association areas send visual information to other brain areas so that an integrated, meaningful interpretation can be made	Involved in planning speech because it analyses the grammatical structure of sentences
Cerebellum	
Involved in the coordination of voluntary movement	
Helps us maintain balance and posture when moving	

4.3.4 Brain labelling and dissection

Key science skills
Apply, evaluate and communicate scientific ideas
*   discuss relevant psychological information, ideas, concepts, theories and models and the connections between them

Develop

MATERIALS

*   coloured pencils or highlighters
*   a semi-frozen sheep's brain
*   scalpel
*   forceps
*   tight-fitting disposable gloves
*   two sheets of newspaper
*   dissecting board

Safety note: Be extremely careful when using the scalpel. Always cut away from your body and hands.

INSTRUCTIONS

PART A

1   In the spaces provided in the captions of Figures 4.8 and 4.9, label the figures as 'left' or 'right' hemisphere.
2   Label the following areas on Figures 4.8 and 4.9.
    *   Motor cortex
    *   Parietal lobe
    *   Somatosensory cortex
    *   Temporal lobe
    *   Frontal lobe
    *   Occipital lobe
    *   Primary visual cortex
    *   Primary auditory cortex
3   Label Broca's area, Wernicke's area and Geschwind's territory on Figure 4.8.
4   Label the following areas on Figure 4.10.
    *   Motor cortex
    *   Frontal lobes
    *   Occipital lobes
    *   Temporal lobes
    *   Parietal lobes
    *   Somatosensory cortex
5   On all three diagrams, use colours to distinguish the lobes. For example: yellow = frontal lobe, blue = temporal lobe, pink = parietal lobe, green = occipital lobe.

Figure 4.8 The _____ hemisphere of the human brain

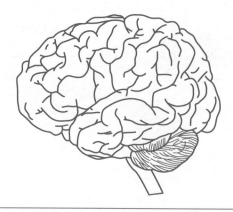

Figure 4.9 The _____ hemisphere of the human brain

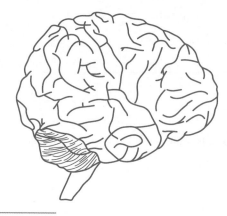

Figure 4.10 View of the human brain from above

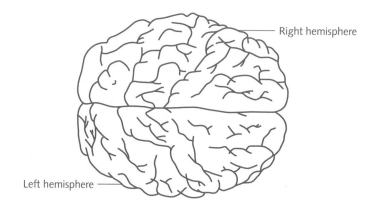

Right hemisphere

Left hemisphere

PART B

1 Wearing the gloves, carefully hold the sheep's brain and identify the areas listed in Part A, step 2.

2 Hold the brain steady on the dissecting board so that you are looking at the top of the cerebral cortex. Carefully cut through the midline of the cerebral cortex and separate the left and right hemispheres.

3 In the space provided below, draw a sketch of one of the hemispheres that shows the inner structures and indicate which hemisphere you are sketching.

4 Sketch in the corpus callosum.

5 When you have finished, follow your teacher's instructions to carefully dispose of all brain parts as required. Thoroughly wash your hands and pack up the dissection board and scalpel as per your teacher's instructions.

Figure 4.11 Sketch of brain dissection showing the corpus callosum

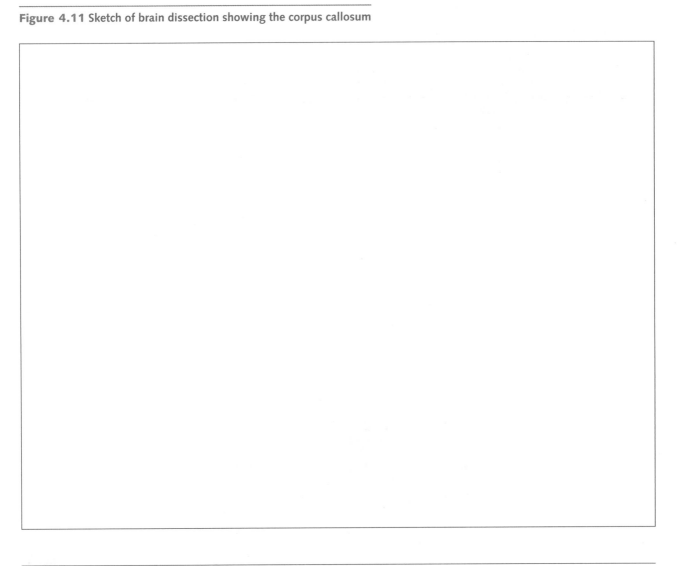

The brain and complex functions

Key knowledge
- the roles of the hindbrain, midbrain and forebrain, including the cerebral cortex, in behaviour and mental processes

4.4.1 Initiation of voluntary movement

Key science skills
Analyse, evaluate and communicate scientific ideas
- discuss relevant psychological information, ideas, concepts, theories and models and the connections between them

Develop

PART A

Use the descriptions from Table 4.13 to label the brain diagram (Figure 4.12).

Table 4.13 Structures involved in the initiation of voluntary movement

I receive information about the body's position in space from other lobes and formulate a plan for motor movement.
I sequence the motor movements.
I transmit motor messages to the skeletal muscles that control voluntary movement.
I provide information about the position and movement of body parts.
I transmit motor information to and from the cerebral cortex to coordinate physical functions such as posture and balance.
I am a relay system that enables motor messages to be transmitted between the cerebral cortex, cerebellum and basal ganglia in order to execute voluntary movements.
I work with the cerebellum to manage smooth coordinated voluntary movement.

Figure 4.12 Areas involved in the initiation of voluntary movement

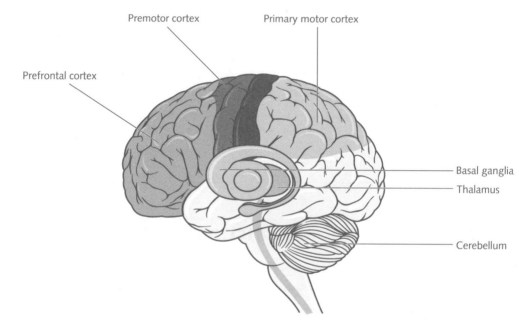

Premotor cortex

Primary motor cortex

Prefrontal cortex

Basal ganglia

Thalamus

Cerebellum

PART B

Imagine you are standing in the kitchen with the fridge to your right. You have decided that you want to open the fridge to find chocolate. Describe a sequence of events that would happen in your brain to initiate your movement over to the fridge. To get to the fridge you will need to walk around the kitchen table.

Develop

4.4.2 Language processing

1 Label the following structures in Table 4.14 on Figure 4.13.

Table 4.14 Terms related to language processing

Broca's area	Wernicke's area
primary auditory cortex	primary motor cortex

Figure 4.13 The process for repeating spoken instructions

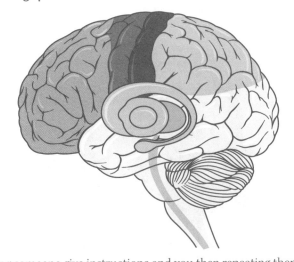

2 Describe the process of hearing someone give instructions and you then repeating them out loud.

- _____
- _____
- _____
- _____

3 Describe the process of repeating instructions you have read from a book.

_____
_____
_____

4.4.3 Brain structures involved in the regulation of emotions

Key science skills

Analyse, evaluate and communicate scientific ideas
- discuss relevant psychological information, ideas, concepts, theories and models and the connections between them

Develop

1 Find and label the following structures in Table 4.15 on Figure 4.14.

Table 4.15 Terms related to the regulation of emotions

thalamus	hippocampus	basal ganglia
prefrontal cortex	hypothalamus	amygdala

Figure 4.14 Structures that assist in regulating emotions

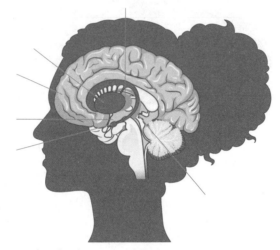

2 Circle those structures that are considered to be a part of the limbic system.

4.4.4 Regulation of emotions: media analysis

Key science skills

Analyse, evaluate and communicate scientific ideas
- discuss relevant psychological information, ideas, concepts, theories and models and the connections between them
- critically evaluate and interpret a range of scientific and media texts (including journal articles, mass media communications, opinions, policy documents and reports in the public domain), processes, claims and conclusions related to psychology by considering the quality of available evidence

Develop

Read the following article and answer the questions that follow.

Why people with anxiety and other mood disorders struggle to manage their emotions

Leanne Rowlands
PhD Researcher in Neuropsychology, Bangor University
January 26, 2019

Regulating our emotions is something we all do, every day of our lives. This psychological process means that we can manage how we feel and express emotions in the face of whatever situation may arise. But some people cannot regulate their emotions effectively, and so experience difficult and intense feelings, often partaking in behaviours such as self-harm, using alcohol, and over-eating to try to escape them.

There are several strategies that we use to regulate emotions – for example, reappraisal (changing how you feel about something) and attentional deployment (redirecting your attention away from something). Underlying neural systems in the brain's prefrontal cortex are responsible for these strategies. However, dysfunction of these neural mechanisms can mean that a person is unable to manage their emotions effectively.

There is not one simple pathway that causes this dysfunction.

In anxiety disorders, dysfunction of the brain's emotional systems is related to emotional responses being of a much higher intensity than usual, along with an increased perception of threat and a negative view of the world.

In the brains of those with anxiety disorders, the system supporting the reappraisal does not work as effectively. Parts of the prefrontal cortex show less activation when this strategy is used, compared to non-anxious people. In fact, the higher the levels of anxiety symptoms, the less activation is seen in these brain areas. This means that the more intense the symptoms, the less they are able to reappraise.

Similarly, those with major depressive disorder (MDD) struggle to use cognitive control to manage negative emotions and decreased emotional intensity. This is due to neurobiological differences, such as decreased density of grey matter, and reduced volume in the brain's prefrontal cortex. During emotion regulation tasks, people who have depression show less brain activation and metabolism in this area.

There is little doubt that people have different abilities in using different regulation strategies. But for some they simply don't work as well. It's possible that people with anxiety disorders find reappraisal a less effective strategy because their attentional bias means they involuntarily pay more attention towards negative and threatening information. This can stop them from being able to come up with more positive meanings for a situation – a key aspect of reappraisal.

It's possible that reappraisal doesn't work as well for people with mood disorders either. Cognitive biases can lead people with MDD to interpret situations as being more negative and make it difficult to think more positive thoughts.

It's important to note that mood disorders don't just come from neural abnormalities. The research suggests that a combination of brain physiology, psychological and environmental factors are what contributes to the disorders, and their maintenance.

While researchers are pursing promising new treatments, simple actions can help people loosen the influence of negative thoughts and emotions on mood. Positive activities like expressing gratitude, sharing kindness, and reflecting on character strengths really do help.

'Why people with anxiety and other mood disorders struggle to manage their emotions' by Leanne Rowland, PhD Researcher in Neuropsychology, Bangor University. The Conversation, January 26, 2019. https://theconversation.com/why-people-with-anxiety-and-other-mood-disorders-struggle-to-manage-their-emotions-106865. Adapted with permission from Leanne Rowland.

Questions

1 People who have difficulty regulating emotions demonstrate increased maladaptive behaviours such as:

_____

_____

2 What is reappraisal? Provide an example.

_____

_____

_____

3 What is attentional deployment? Provide an example.

_____

_____

4 People with anxiety show an/a _____ in prefrontal activity.

5 What structural changes can be seen in people who suffer from major depressive disorder? (MDD)

6 What is the role of cognitive bias in emotion regulation for people with anxiety or MDD?

7 What strategies are helpful in managing mood disorders?

4.4.5 Brain damage case studies

Key science skills
Analyse, evaluate and communicate scientific ideas
• discuss relevant psychological information, ideas, concepts, theories and models and the connections between them

Develop

Read the case studies and complete the questions that follow.

CASE STUDY 1

Lia was riding her bike down a hill to school when she realised her brakes were faulty. At the bottom of the hill, she hit a tree and flew off the bike, falling to the ground and hitting the back of her head on the gutter. After her accident, Lia had great difficulties walking in a smooth, coordinated way. She had trouble maintaining her balance, which often left her feeling dizzy.

Questions

1 Name the condition it is likely Lia is experiencing.

2 Name the brain structure to which Lia sustained damage.

3 Is this brain structure in the midbrain, hindbrain or forebrain?

CASE STUDY 2

Leesa was a healthy 22-year-old woman who ate a balanced diet and enjoyed working out at the gym two or three times each week. She maintained a healthy weight range, as measured by her BMI (body mass index). One day Leesa suffered a severe stroke. For some time after her stroke Leesa had difficulties controlling her hunger and thirst. She began overeating, could not seem to control her appetite and would often binge eat. As a result, she gained 10 kilograms within two months.

Questions

1 What is the brain structure that has been damaged by Leesa's stroke?

2 Identify other changes in Leesa's behaviour that could occur as a result of damage to this area of her brain.

CASE STUDY 3

Cybil was at her school's annual carnival. She was trying all of the scary rides with her best friend, Beth. The girls were on a ride when it malfunctioned, and they were thrown onto the hard concrete below. Both had to be rushed to hospital. Cybil sustained life-threatening brain injuries and was placed on life support because her heart rate and breathing were not self-regulating. Beth sustained non-critical injuries, although she experienced difficulties with her vision.

Questions

1 What part of Cybil's brain was most likely to have been damaged during the accident?

2 Why were Cybil's breathing and heart rate no longer self-regulating?

3 What part of Beth's brain was most likely to have been damaged during the accident?

CASE STUDY 4

Bill was cleaning his roof gutters when he slipped and fell to the ground. He was taken to hospital in an ambulance and as soon as his wife was called, she rushed to the hospital. When she arrived, Bill did not recognise her as his wife even though they had been married for 25 years. The doctors asked Bill to describe the physical features of his wife's face. Bill was able to do this: he described his wife as having short brown hair, blue eyes, thin lips and high cheekbones, and all these details were correct.

Questions

1 What area of Bill's brain was likely to be affected by the accident?

2 Describe three other symptoms Bill may experience as a result of the brain trauma.

CASE STUDY 5

Anthony was riding a dirt bike on his farm when he hit a pothole and fell off. Because he was not wearing a helmet, he sustained a serious head injury. After the accident, Anthony was unable to produce clear and fluent speech. For example, if he wanted to say, 'What are we having for dinner tonight?', he might instead say 'What … have … dinner … tonight?'

Questions

1 Name the area of the brain to which Anthony has most likely sustained damage.

2 What is the name of the condition doctors are likely to have diagnosed in Anthony?

3 What lobe of the cerebral cortex has been affected?

4 Explain three other functions of the lobe that has been damaged and name some other subsequent difficulties Anthony may experience.

4.4.6 Evaluation of research

Key science skills

Develop aims and questions, formulate hypotheses and make predictions
- identify independent, dependent, and controlled variables in controlled experiments
- formulate hypotheses to focus investigation

Comply with safety and ethical guidelines
- demonstrate ethical conduct and apply ethical guidelines when undertaking and reporting investigations

Construct evidence-based arguments and draw conclusions
- evaluate data to determine the degree to which the evidence supports or refutes the initial prediction or hypothesis

Develop

Read the research scenario and complete the questions that follow.

Neurologist Dr Robert Geralovski was interested in finding out who is more likely to have language structures located in the non-specialised right hemisphere. He believed there might be a relationship between 'handedness' (that is, being left- or right-handed) and language hemispheric specialisation. He proposed that because of the brain's contralateral organisation, there would be a large portion of left-handers who had brain structures responsible for language located in their right hemispheres.

To carry out this study, Dr Geralovski recruited 20 left-handed participants and 20 right-handed participants. He selected his participants by choosing the first 40 people who responded to an advertisement he placed in his local newspaper. He followed informed consent procedures by enabling participants to read a document outlining the potential risks of the experiment, as well as explaining the necessity for participants to remain in the experiment for the entire length of the study. Dr Geralovski separated his participants into two groups. Group 1 comprised the left-handed participants and Group 2 comprised the right-handed participants.

Once the participants had been allocated to groups, he began the experiment by using a technique known as the Wada Test. The Wada Test involves anaesthetising one hemisphere of the brain. When this is done a person temporarily loses the functions of that hemisphere.

Dr Geralovski applied the following procedure to both groups. He used the Wada test to anaesthetise each participant's right hemisphere. He then asked the participant to respond verbally to three simple questions. He recorded the responses and analysed them for accuracy of verbal communication.

The results of his study are listed in Table 4.16.

Dr Geralovski found that left-handed participants who had had their right hemisphere anaesthetised performed almost as well on the verbal questions as their right-handed counterparts. This suggested that there was no relationship between handedness and language dominance. That is, a majority of both left-handed and right-handed participants accurately responded to the questions, indicating that the language centre was in the left (active) hemisphere.

Table 4.16 Left- and right-handed participants who accurately responded to questions verbally

Group	Participants who accurately responded to questions verbally
Group 1: Left-handers	84%
Group 2: Right-handers	89%

1   What was the aim of this experiment?

2   Write a possible hypothesis for this experiment.

3   Identify the independent variable.

4   Identify the dependent variable.

5   Identify a possible extraneous variable that may have affected the results of the experiment.

6   What sampling procedure did Dr Geralovski use to recruit his participants?

7   Identify the experimental and control groups.

8   What type of data was collected from this research?

9   Was Dr Geralovski's research ethical? Which principles did he follow (or not follow)?

10 Write a conclusion for this experiment.

Chapter 4 summary

Photocopy (increasing the size) pages 116–118 onto A3 paper.
Cut out each triangle and then match the term with the definition.
You should make a hexagon shape.

You may need to do extra research for some of the terms.

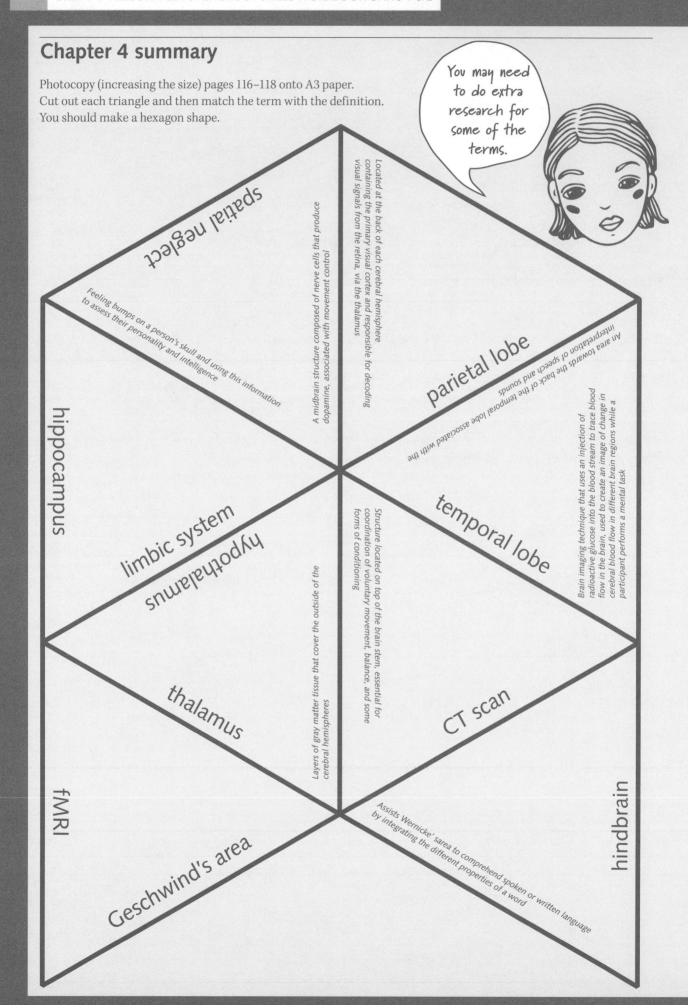

spatial neglect

Located at the back of each cerebral hemisphere containing the primary visual cortex and responsible for decoding visual signals from the retina, via the thalamus

A midbrain structure composed of nerve cells that produce dopamine, associated with movement control

parietal lobe

An area towards the back of the temporal lobe associated with the interpretation of speech and sounds

Feeling bumps on a person's skull and using this information to assess their personality and intelligence

hippocampus

limbic system

hypothalamus

temporal lobe

Brain imaging technique that uses an injection of radioactive glucose into the blood stream to trace blood flow in the brain, used to create an image of change in cerebral blood flow in different brain regions while a participant performs a mental task

Structure located on top of the brain stem, essential for coordination of voluntary movement, balance, and some forms of conditioning

thalamus

Layers of gray matter tissue that cover the outside of the cerebral hemispheres

CT scan

fMRI

Geschwind's area

Assists Wernicke's area to comprehend spoken or written language by integrating the different properties of a word

hindbrain

9780170465045

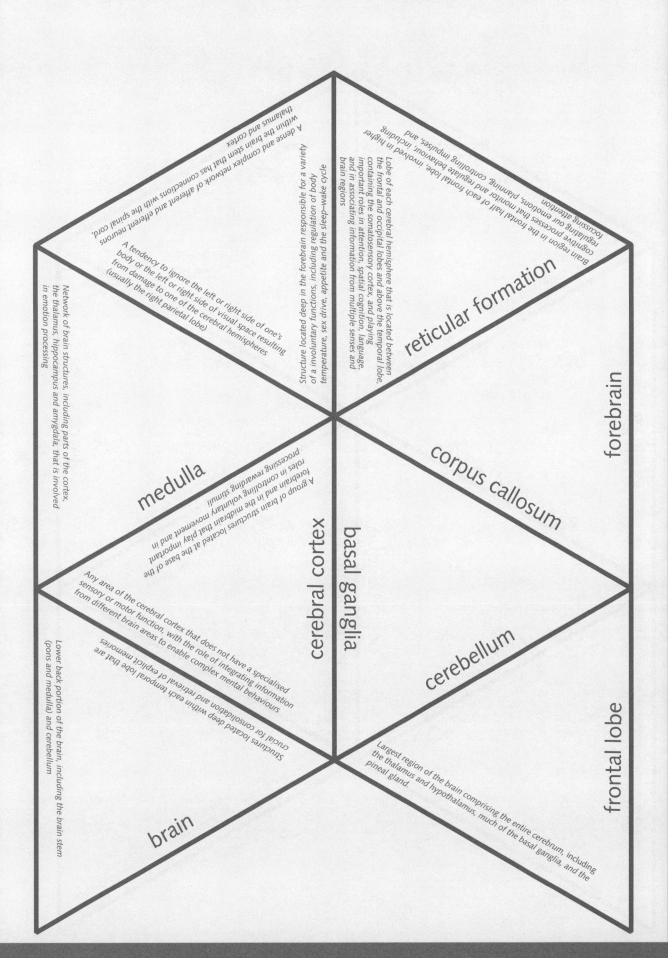

A dense and complex network of afferent and efferent neurons within the brain stem that has connections with the spinal cord, thalamus and cortex

Brain region in the frontal lobe, involved in higher cognitive processes that monitor and regulate behaviour, including regulating our emotions, planning, controlling impulses, and focussing attention

reticular formation

Lobe of each cerebral hemisphere that is located between the frontal and occipital lobes and above the temporal lobe, containing the somatosensory cortex, and playing important roles in attention, spatial cognition, language, and in associating information from multiple senses and brain regions

A tendency to ignore the left or right side of one's body or the left or right side of visual space resulting from damage to one of the cerebral hemispheres (usually the right parietal lobe)

Structure located deep in the forebrain responsible for a variety of involuntary functions, including regulation of body temperature, sex drive, appetite and the sleep–wake cycle

forebrain

corpus callosum

medulla

A group of brain structures located at the base of the forebrain and in the midbrain that play important roles in controlling voluntary movement and in processing rewarding stimuli

Network of brain structures, including parts of the cortex, the thalamus, hippocampus and amygdala, that is involved in emotion processing

cerebral cortex

basal ganglia

cerebellum

Any area of the cerebral cortex that does not have a specialised sensory or motor function, with the role of integrating information from different brain areas to enable complex mental behaviours

Structures located deep within each temporal lobe that are crucial for consolidation and retrieval of explicit memories

Lower back portion of the brain, including the brain stem (pons and medulla) and cerebellum

Largest region of the brain comprising the entire cerebrum, including the thalamus and hypothalamus, much of the basal ganglia, and the pineal gland.

frontal lobe

brain

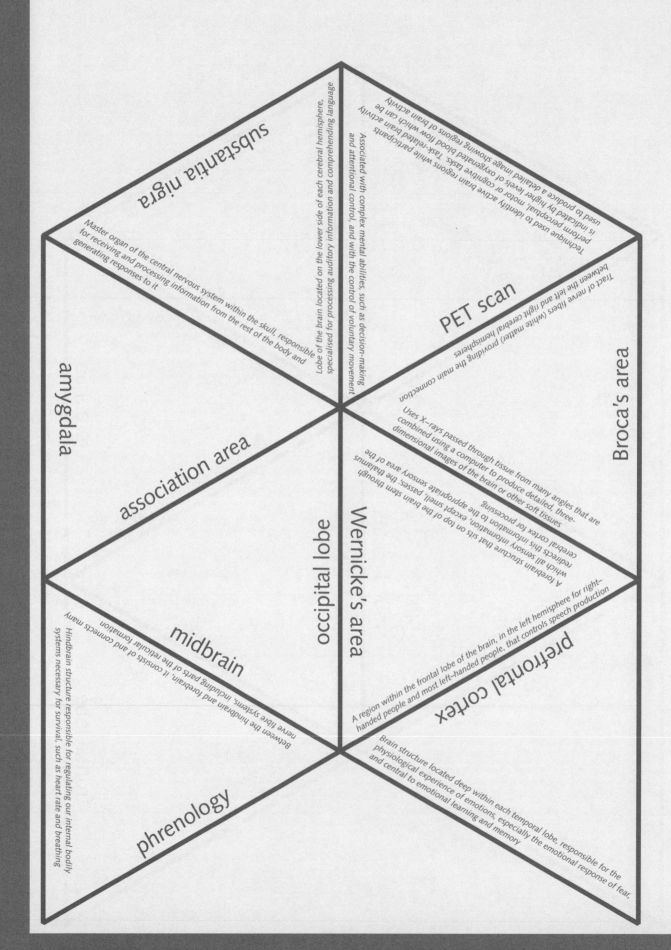

substantia nigra

Lobe of the brain located on the lower side of each cerebral hemisphere, specialised for processing auditory information and comprehending language

Associated with complex mental abilities, such as decision-making and attentional control, and with the control of voluntary movement

Technique used to identify active brain regions while participants perform perceptual, motor or cognitive tasks. Task-related brain activity is indicated by higher levels of oxygenated blood flow which can be used to produce a detailed image showing regions of brain activity

Master organ of the central nervous system within the skull, responsible for receiving and processing information from the rest of the body and generating responses to it

PET scan

Tract of nerve fibres (white matter) providing the main connection between the left and right cerebral hemispheres

amygdala

Broca's area

association area

Uses X-rays passed through tissue from many angles that are combined using a computer to produce detailed, three-dimensional images of the brain or other soft tissues

A forebrain structure that sits on top of the brain stem through which all sensory information, except smell, passes; the thalamus redirects this information to the appropriate sensory area of the cerebral cortex for processing

occipital lobe

Wernicke's area

midbrain

prefrontal cortex

Between the hindbrain and forebrain, it consists of and connects many nerve fibre systems, including parts of the reticular formation

A region within the frontal lobe of the brain, in the left hemisphere for right-handed people and most left-handed people, that controls speech production

phrenology

Hindbrain structure responsible for regulating our internal bodily systems necessary for survival, such as heart rate and breathing

Brain structure located deep within each temporal lobe, responsible for the physiological experience of emotions, especially the emotional response of fear, and central to emotional learning and memory

9780170465045

Brain plasticity and brain injury

5

5.1 Neuroplasticity: rewiring the brain

Key knowledge
- the capacity of the brain to change in response to experience and brain trauma including factors influencing neuroplasticity and ways to maintain and/or maximise brain functioning

5.1.1 Parts of a neuron

Key science skills
Analyse, evaluate and communicate psychological ideas
- discuss relevant psychological information, ideas, concepts, theories and models and the connections between them

Develop

PART A

1   Use the terms in Table 5.1 to label the parts of a basic neuron in Figure 5.1.

Table 5.1 The parts of a neuron

axon	dendrite	soma	actin filaments
microtubule	myelin sheath	axon terminals	dendritic spine

2   Mark with an X the point in Figure 5.1 at which a neurotransmitter would be released.

This activity will build on your understanding of the brain from Chapter 4.

Figure 5.1 A basic neuron showing two types of cytoskeleton, microtubules and actin filaments, both involved in giving the neuron structure and plasticity.

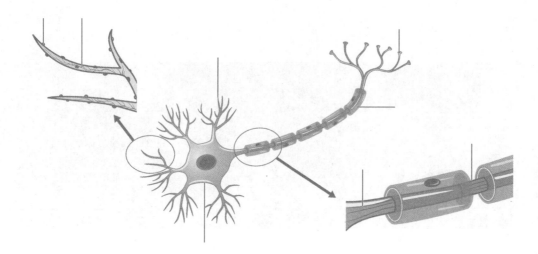

PART B

Use the terms in Table 5.2 to fill in the blanks next to the numbered statements that follow.

Some terms may be used more than once.

Table 5.2 Terms associated with neurons

soma	neurotransmitters	cytoskeleton	dendrites	myelin
synaptic gap	receptor site	axon	axon terminals	nodes of Ranvier
electrical	neuron	neuromodulator	learning	spines

1   Transmits information to the soma _____

2   A single, long, thin fibre that carries information away from the soma _____

3   The cell body that receives incoming information from its dendrites _____

4   Chemicals found in the synapses between neurons _____

5   Branching fibres at the end of the axon _____

6   The part of a dendrite's surface that receives a neurotransmitter _____

7   The white, fatty substance that protects the axon and helps transmit the neural impulse along the axon _____

8   Gaps in the myelin sheath _____

9   Sends the neural impulse across the synapse to a dendrite _____

10 They either excite or inhibit the postsynaptic neuron _____

11 Branches (neuron fibres) extending from the soma that receive information from other cells _____

12 The minute space between one cell's axon terminal and another cell's dendrite _____

13 Their release can excite or inhibit the activity of a neighbouring neuron _____

14 A chemical that is released by a neuron which influences (modulates) the transmission of signals between neurons _____

15 Proteins that provide support and structure to a neuron _____

PART C

Use the terms in Table 5.2 to fill in the blanks in the following paragraphs.

Some terms may be used more than once.

A _____ is an individual nerve cell specialised to receive, transmit and process information. Billions of these nerve cells are interconnected to form the nervous system so that information can be transmitted around the body. All neurons share a basic structure. The _____, or cell body, processes information brought to it by a number of projecting branches, or _____, which receive information from other neurons. At the other end of the cell body, a single _____ transmits the information away from the cell body. At this stage, the information exists in the form of an _____ impulse.

Some axons are covered by a white, fatty substance called _____, which forms a protective layer around the axon and insulates it. Small gaps that occur at regular intervals, called _____ allow the impulse to jump from gap to gap, so they increase the speed of transmission. When the impulse reaches the end of the axon, the axon branches out into a number of _____ that release chemical _____ into the _____ between the dendrites and axon terminals of neurons. These chemicals carry the messages to the _____ on the dendrite of a neighbouring neuron. The dendrite then passes the information to its soma and the process of neural transmission continues. Dendritic _____ are a mushroom-shaped outgrowth along the dendrite of a neuron. Research suggests that neurons develop dendritic _____ as part of the _____ process. The neuronal _____ plays a crucial role in maintaining cell shape and function. Deformities of the _____ have been reported to be associated with neurodegenerative diseases.

5.1.2 Brain development

> **Key science skills**
>
> Analyse and evaluate data and investigation methods
> - process quantitative data using appropriate mathematical relationships and units, including calculations of percentages, percentage change and measures of central tendencies (mean, median, mode), and demonstrate an understanding of standard deviation as a measure of variability
> - identify and analyse experimental data qualitatively, applying where appropriate concepts of: accuracy, precision, repeatability, reproducibility and validity; errors; and certainty in data, including effects of sample size on the quality of data obtained
>
> Analyse, evaluate and communicate scientific ideas
> - discuss relevant psychological information, ideas, concepts, theories and models, and the connections between them
>
> Develop

PART A

Indicate whether each statement is true or false by placing a tick in the correct column in Table 5.3. If the statement is false, underline the incorrect words and identify the reason why it is incorrect in the 'Correction' column.

Statement 1 has been done for you as an example of what to do.

Table 5.3 Processes associated with brain development

Description	True	False	Correction
The brain begins to develop during the <u>5th week of gestation</u>.		✔	3rd week of gestation
By the time an individual reaches adulthood, their brain is composed of more than 100 billion neurons.			
Cell elaboration involves the newly formed neurons travelling to their final location within the nervous system; the location determines what their function will be.			
Cell production, cell migration and cell elaboration only occur during gestation.			
Brain cells dividing and multiplying refers to cell proliferation.			
The role of the myelin sheath is to protect the neuron's axon, and this increases the speed at which the electrical impulse travels.			
The parts of the brain that contain myelinated neurons are referred to as grey matter.			
To assist brain development, the infant brain has excess neurons during the first three years of life.			
Synaptic pruning is a process by which synaptic connections are removed so that neuronal transmission is as efficient as possible.			
During infancy, synaptic pruning possibly reduces the number of synapses in the human brain by approximately 40%.			

PART B

Use the terms in Table 5.4 to fill in the blanks in the following paragraphs. Each word is used only once.

Table 5.4 Brain development terms

apoptosis	strengthen	white	weak	pruning	myelination
axon	grey	complex	neurons	adolescence	adulthood
excess	synapses	myelinated	three	learning	efficient
spine	cytoskeleton				

Myelination of _____ in the human brain begins during foetal development, and research suggests that it continues through adolescence and into early _____. This is the process where myelin, a white, fatty substance, coats the _____ of a neuron. Neurons in the cortex that are unmyelinated are referred to as _____ matter, whereas parts of the brain that contain myelinated neurons are referred to as _____ matter. As the brain becomes more _____ throughout development, information is transmitted more efficiently between areas. To assist brain development, the infant brain has _____ neurons during the first _____ years of life. This is why _____ new things is easier in infancy compared to other lifespan stages. Due to the oversupply of neurons in the infant brain, not all neurons will be activated. The neurons and synapses that are repetitively activated _____, but those that are rarely used remain _____ and are more likely to be eliminated through the process of _____.

During adolescence, _____ continues and the brain becomes more developed; hence the adolescent is able to perform more _____ behaviours than an infant. Synaptic _____ also continues during _____, where synaptic connections are removed so that neuronal transmission is as _____ as possible. This reduction in synaptic density possibly reduces the number of _____ by approximately 40 per cent. Developmentally, dendritic _____ plasticity is prominent until adolescence and then drops in adulthood to low levels.

PART C

Draw a line to match the terms on the right-hand side of Table 5.5 to its description.

Table 5.5 Brain development processes

Neurogenesis	Synapses are formed between neurons so information can be transmitted.
Neural migration	Myelin sheaths coat a neuron's axons.
Neural maturation	Brain cells divide and multiply, forming new neurons.
Synaptogenesis	Brain neurons travel to their final destination, which determines the neuron's function.
Synaptic pruning	Extra, weak or unused synaptic connections are removed.
Myelination	Brain neurons extend their axons and grow dendrites to form synapses.

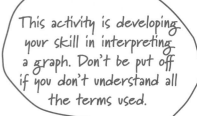

This activity is developing your skill in interpreting a graph. Don't be put off if you don't understand all the terms used.

PART D

Analyse Figure 5.2 and answer the questions that follow.

Figure 5.2 Key brain changes across the life span

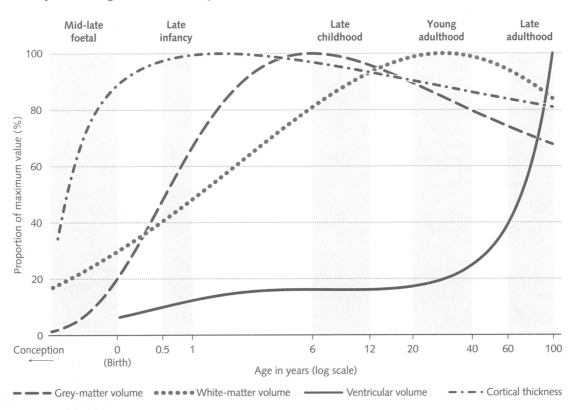

1   Describe the change in ventricular volume from birth to late adulthood.

_____

_____

2 At which stage is cortical thickness at its highest?

_____

3 Account for why grey matter volume decreases after late childhood.

_____
_____
_____

4 What is the purpose of using median values as data for this graph?

_____
_____
_____

5.1.3 Adaptive plasticity

Key science skills

Analyse, evaluate and communicate scientific ideas

- discuss relevant psychological information, ideas, concepts, theories and models, and the connections between them
- critically evaluate and interpret a range of scientific and media texts (including journal articles, mass media communications, opinions, policy documents and reports in the public domain), processes, claims and conclusions related to psychology by considering the quality of available evidence
- acknowledge sources of information and assistance, and use standard scientific referencing conventions

Develop

PART A

Use the terms in Table 5.6 to fill in the blanks in the following paragraphs.

Table 5.6 Adaptive plasticity terms

learning	stimulation	damaged	new	connections	rerouting	undamaged

Adaptive plasticity occurs when the brain alters the _____ between synapses and reorganises neural pathways because of _____ or following injury.

Following brain injury neurons in an _____ brain region can sometimes take over sensory and motor functions that had been performed in _____ areas. This happens after repeated _____ .

Sprouting occurs when _____ connections between neurons are created. _____ occurs when an undamaged neuron that has lost connection with a damaged neuron can make new connections with other active neurons.

PART B

Read the case study of Barbara Arrowsmith-Young and the Arrowsmith Program, then answer the questions in response to the case study.

Try underlining or highlighting the key information.

Arrowsmith Program

Barbara Arrowsmith-Young is the founder of a school for children with learning difficulties, the Arrowsmith School. Her school's teaching strategies are based on scientific research in neuroplasticity, as well as the evidence from her personal successes in overcoming the learning difficulties she experienced.

When Barbara was in Grade 1, she was diagnosed as having a 'mental block'. At that time many learning disabilities were not labelled the way they are today. Today she might have been diagnosed with dyslexia (a reading disorder) and dyscalculia (difficulty learning arithmetic). Barbara had trouble processing language: she was unable to read and write a sentence from left to right, she confused the letter 'b' with 'd', and 'p' with 'q', and read 'saw' as 'was'. Some other deficits Barbara experienced included spatial reasoning, logical thinking, and even the ability to complete simple tasks such as reading an analogue clock. She could not tell the difference between the left and the right hand of the clock because she couldn't understand the relationship between the hands and the numbers on the clock, or distinguish left from right.

Through phenomenal effort, Barbara trained her brain to overcome her learning difficulties by repeatedly practising cognitive exercises that forced her brain to make new neural connections to compensate for her deficiencies. The first technique she developed was to force her brain to process relationships. She did this by reading clocks, starting with a two-handed clock and practising for up to 9 hours per day. Once she mastered this task and was able to read this clock with great speed and accuracy, she began practising with three- and four-handed clocks (with hands for seconds and milliseconds). Once again, through constant repetition and extraordinary effort, she mastered reading these clocks and could do so with great speed and accuracy. This is when she started to see results; she was now able to process spatial relationships.

After her personal successes, Barbara began reflecting on early experiments in neuroplasticity. These had revealed that the brain is not a fixed organ; rather, it is malleable and can change drastically in response to growth and experience throughout a person's whole life. Using this research and her own experiences, she developed the Arrowsmith Program for students with learning difficulties.

The Arrowsmith School aims to improve the cognitive capacity of students with learning disabilities. Students from all over the world attend the school, which started in Toronto, Canada. The syllabus does not involve traditional Maths, English and Science subjects; in contrast, students at this school focus on persistent and rigorous cognitive exercises. In ordinary schools, students with learning difficulties will typically be given a modified learning program, with compensatory activities or a reduction in the amount and complexity of learning tasks in the classroom. For example, if a student has been diagnosed with dysgraphia (a learning disability that affects writing ability), this student may be given more writing time in exams or permission to use a computer to compensate for their difficulty. The Arrowsmith Program takes a different approach, and instead encourages students to strengthen weak cognitive connections that are enabling their dysfunctions. By challenging and strengthening these connections students begin to overcome their difficulties so they no longer require modified programs. Students who have attended the Arrowsmith School and completed the cognitive exercises have shown improvement in their performance on tasks such as reading, writing, mathematics and comprehension of conceptual material. This has all been made possible because of the plasticity of their brains.

You can read more about Barbara's story in Chapter 2 of Norman Doidge's 2007 book, *The Brain That Changes Itself*, or in Barbara's own book, *The Woman Who Changed Her Own Brain*.

Sources

Arrowsmith School. (n.d.). Background. https://arrowsmithschool.org/background/
Barbara Arrowsmith Young. (n.d.). Bio. https://barbaraarrowsmithyoung.com/bio/
Doidge, N. (2007). The brain that changes itself. Viking Penguin: USA

Questions

1 What is adaptive plasticity?

2 How has adaptive plasticity helped Barbara Arrowsmith-Young overcome her learning difficulties?

3 Describe one of the cognitive techniques Barbara used to overcome her learning disability.

_____

_____

4 Suggest a reason why the Arrowsmith School is so successful in helping children overcome learning difficulties.

_____

_____

_____

5 Go online and find another story that highlights the brain's capacity to rewire itself and overcome damage and/or disability. Write a summary of this story below. Make sure that you provide the source of your story according to APA reference conventions.

_____

_____

_____

_____

_____

_____

_____

_____

5.2   A brain injury

Key science knowledge
- the impact of an acquired brain injury on a person's biological, psychological and social functioning

5.2.1 Acquired brain injury

Key science skills
Analyse, evaluate and communicate scientific ideas
- discuss relevant psychological information, ideas, concepts, theories and models, and the connections between them

Develop

Use the terms in Table 5.7 to fill in the blanks in the following paragraphs.

Table 5.7 Brain injury terms

temporary	inflammation	oxygen	traumatic	deteriorating	contact
external	slowly	emotional	alcohol	total	birth
haemorrhage					

Acquired brain injury (ABI) refers to any type of brain damage that happens after _____.
When a person has an ABI a portion of their brain cells are destroyed or are _____.

Acquired brain injuries can occur suddenly or they can develop over time. They can be _____
or permanent and may result in partial or _____ cognitive, physical, _____
or motor impairment.

Acquired brain injuries may occur suddenly because of injury by an _____
force such as a traumatic blow to the head that damages brain tissue or structures. This type of ABI is known
as a _____ brain injury (TBI). They may result from events such as a fall, an assault,

a car accident, shaken baby syndrome or during _____ sports. A TBI is not the same as a head injury, since a person can experience an impact to the head and sustain damage to the face, scalp and skull without necessarily injuring their brain.

Brain injury that occurs _____ over time because of internal factors is referred to as a non-traumatic brain injury (NTBI). Causes of NTBI include lack of _____ to the brain (known as hypoxia or anoxia) that injures brain tissue, exposure to toxins (such as excessive and prolonged _____ and drug use), infectious disease that causes _____ to the brain covering or the brain tissue itself, tumors that damage surrounding brain tissue and structures, a stroke as a result of a _____, or blockage to blood vessels that supply blood to the brain. Degenerative neurological diseases that cause abnormal changes to brain cells, such as Alzheimer's disease, multiple sclerosis or Parkinson's disease can also cause brain injury.

5.2.2 Areas of the cerebral cortex: the effect of brain damage

Key science skills
Analyse, evaluate and communicate scientific ideas
- discuss relevant psychological information, ideas, concepts, theories and models, and the connections between them

Develop

MATERIALS

- scissors
- glue

INSTRUCTIONS

Table 5.8 describes some of the symptoms that can occur due to damage to specific parts of the cerebral cortex.
1 Cut out each symptom of brain damage in Table 5.8.
2 Paste each symptom into the box that is linked to the correct lobe of the cerebral cortex in Figure 5.3.

If you need, use Chapter 4 to help you out.

Table 5.8 Some symptoms of brain damage

Inability to concentrate	Loss of vision
Paralysis of a body part	Impaired spatial awareness
Difficulty recognising visual material such as shapes or objects	Inability to differentiate between hot and cold temperatures
Inability to feel pain	Inability to hear
Memory loss	Problems with working memory
Difficulty regulating emotions	Difficulty judging distance
Difficulty recognising words	Difficulty expressing spontaneous emotions through facial expressions
Personality change	Difficulty recognising faces

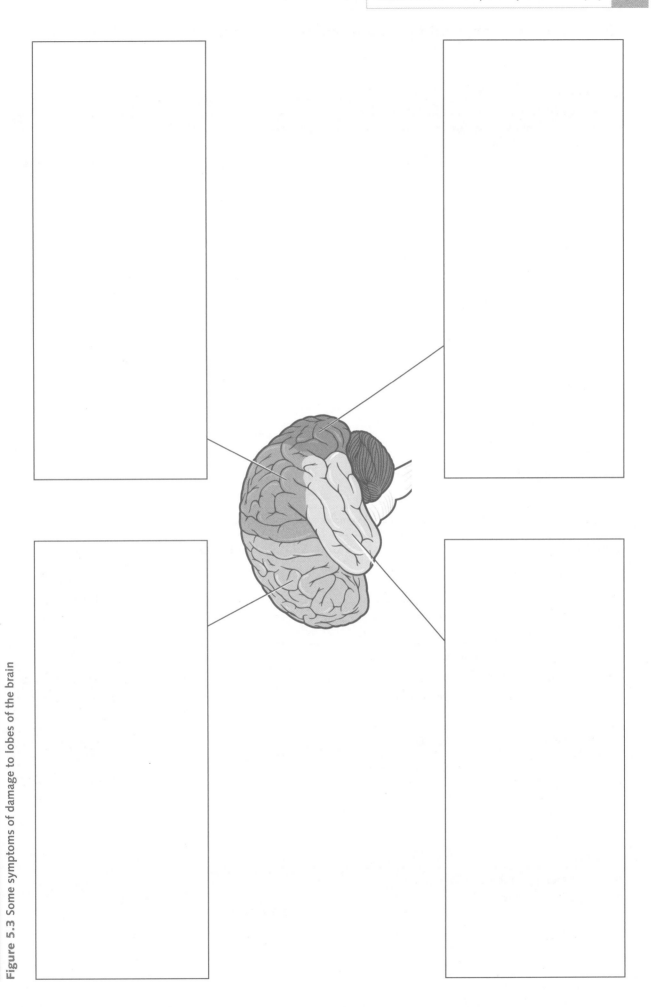

Figure 5.3 Some symptoms of damage to lobes of the brain

5.2.3 Biological, psychological and social dysfunctions associated with the cerebral cortex

Key science skills
Analyse, evaluate and communicate scientific ideas
• discuss relevant psychological information, ideas, concepts, theories and models, and the connections between them

Develop

PART A

Answer the questions below.
What is meant by:

a biological functioning?

_____

_____

b psychological functioning?

_____

_____

c social functioning?

_____

_____

PART B

Read each symptom described in Table 5.9 and tick the appropriate box to indicate whether it is a biological, psychological or social dysfunction.

Table 5.9 Some biological, psychological and social symptoms of brain damage

Symptom	Biological	Psychological	Social
Having difficulty walking in a smooth, coordinated motion	✔		
Having difficulties problem solving; for example, solving a maths equation			
Laughing inappropriately when having a conversation with a friend about the death of a loved one			
Experiencing involuntary muscle spasms in your arms and legs			
Having difficulty judging distance when trying to put a coffee mug on a table			
Eating two ice-cream flavours (chocolate and vanilla) but being unable to taste the difference between the two flavours			
Forgetting your 16th birthday party, which was two weeks ago			
Being unable to have a conversation with your friend because of language difficulties			
Experiencing difficulty seeing clearly			
Being unable to read a Facebook message from your friend inviting you to her birthday party			

The first one has been completed as an example of what to do.

5.2.4 Spatial neglect

Key science skills
Analyse, evaluate and communicate scientific ideas
• discuss relevant psychological information, ideas, concepts, theories and models, and the connections between them

Develop

1 Imagine you have damage to the rear of your left parietal lobe. Draw in the space on the right what you would see in the left column of Figure 5.4.

Figure 5.4 Features of spatial neglect

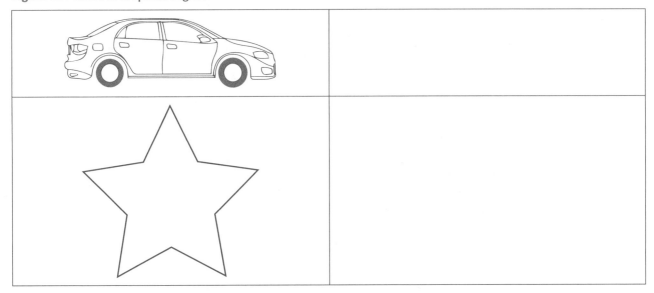

2 Locate and circle only the small stars without marking the large stars or letters on Figure 5.5.

Figure 5.5 Features of spatial neglect: an example of a clinical tool

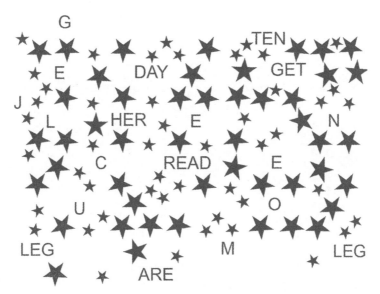

3 Justify your annotations using your understanding of spatial neglect.

5.2.5 Features of Broca's aphasia and Wernicke's aphasia

Key science skills
Analyse, evaluate and communicate scientific ideas
• discuss relevant psychological information, ideas, concepts, theories and models, and the connections between them

Develop

PART A

Read each symptom described in the first column of Table 5.10 and tick the appropriate box to indicate whether it is a feature of Broca's aphasia or Wernicke's aphasia.

Table 5.10 Features of Broca's aphasia and Wernicke's aphasia

Feature	Broca's aphasia	Wernicke's aphasia
Person unable to produce clear, fluent speech that others can understand.		
Person is often unaware of their ability to communicate effectively.		
Person unable to understand language (written and spoken).		
Person can understand language.		
Person is usually aware of their inability to communicate effectively.		
Person's speech is slow and slurred and words are not properly formed.		
Person puts words together that are not grammatically correct and creates sentences that do not make sense.		
Person's speech is clear and words are properly formed.		

PART B

Use the information from your completed Table 5.10 to create a Venn diagram to show how Broca's aphasia and Wernicke's aphasia are similar and different.

Figure 5.6 Venn diagram

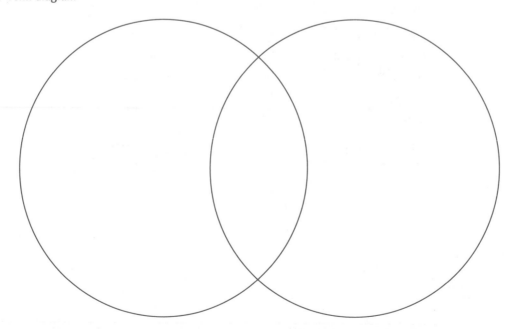

5.2.6 Brain damage case studies

Key science skills

Analyse, evaluate and communicate scientific ideas
- discuss relevant psychological information, ideas, concepts, theories and models, and the connections between them

Develop

Read the case studies below and complete the questions that follow each one.

CASE STUDY 1

Uncle Bob was hospitalised after he fell off a ladder and hit his head. Before this accident he was able to do mental arithmetic faster than anyone in the family, and he was very good at making sensible decisions and planning ahead. However, after his accident he found even simple arithmetic problems difficult to solve, his decision-making was poor and he seemed unable to plan for the future.

Questions

1 What lobe of Uncle Bob's brain was probably damaged in the fall?

2 Identify changes in other behaviours that Uncle Bob might have experienced because of damage to this lobe.

CASE STUDY 2

One day when Danilo was jogging along the riverside, he was caught in an unexpected storm. The winds were fierce and, as Danilo ran under a tree, a branch fell and hit him on the head. Although he was not knocked unconscious, Danilo felt stunned by the blow. That night he had a slight headache, so he went to bed early. When he got out of bed the next morning his left leg felt numb and he had difficulty controlling its movement.

Questions

1 Name the lobes of Danilo's brain that were probably damaged.

2 Name the specific area of the damaged lobe that caused the numbing sensation in Danilo's leg.

3 Name the specific area of the damaged lobe that impaired Danilo's ability to direct the movement of his leg.

4 Name the hemisphere in which this damage occurred.

CASE STUDY 3

When Tim was 10 years old he fell down a flight of stairs. Prior to his fall, Tim was a placid, even-tempered boy who liked to spend his time reading novels or playing football with his many friends. Six months after the fall, his parents took him to their doctor because they were worried about his behaviour. Tim showed signs of hyperactivity and, because he was often angry and aggressive, none of his former friends would play with him. He also seemed to have difficulty kicking the football with his right leg.

Questions

1 Name the lobes of Tim's brain that were probably damaged in the fall.

2 Offer a possible explanation for Tim's inability to kick the football with his right leg.

CASE STUDY 4

During a rugby union game, Vasil was involved in a number of tackles. When he stood up after one tackle, Vasil felt dizzy and his vision was blurred. After a few minutes, his dizziness subsided but he still had difficulty seeing clearly. The next morning when he awoke, he experienced black spots in his vision.

Questions

1 Name the specific area of the lobes of Vasil's brain that was probably damaged in the tackle.

2 Give a possible explanation of why damage to this area of the brain would result in his visual impairment.

CASE STUDY 5

Katrina was driving her car on the freeway when she was involved in a car accident. Fortunately, Katrina survived the accident; however, she did sustain an acquired brain injury (ABI). Symptoms of her ABI included the inability to notice things in her left visual field. She was unable to attend to items on the left side of the table she was sitting at, unable to read the left side of a clock to tell the time and often only ate food on the right side of her plate.

Questions

1 What area of the brain did Katrina sustain damage to?

2 What is the name of the condition Katrina is suffering from?

3 Why can Katrina only pay attention to the right side of her visual field?

5.3 # Neurological disorders

Key knowledge
- the contribution of contemporary research to the understanding of neurological disorders

5.3.1 Parkinson's disease

Key science skills
Analyse, evaluate and communicate scientific ideas
- discuss relevant psychological information, ideas, concepts, theories and models, and the connections between them

Develop

1   Research Parkinson's disease and write down the details of the disease in the graphic organiser in Figure 5.7.

2   Under 'Parkinson's disease' provide an overview of the disease.

3   In the other boxes include descriptions of the types of symptoms, and the possible causes and treatments for sufferers of Parkinson's disease.

4   Use your textbook and the Parkinson's Australia website to help with this task. (Enter 'Parkinson's Australia' into a search engine.)

Figure 5.7 A summary of Parkinson's disease

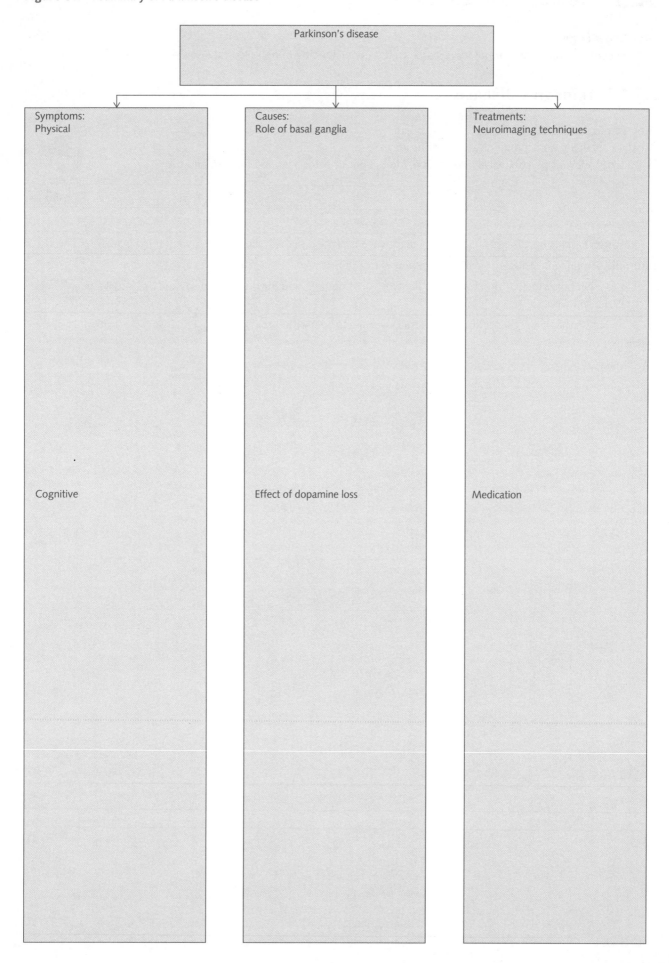

Parkinson's disease

Symptoms:
Physical

Cognitive

Causes:
Role of basal ganglia

Effect of dopamine loss

Treatments:
Neuroimaging techniques

Medication

5.3.2 Neuroimaging techniques

Key science skills

Plan and conduct investigations

- design and conduct investigations; select and use methods appropriate to the investigation, including consideration of sampling technique (random and stratified) and size to achieve representativeness, and consideration of equipment and procedures, taking into account potential sources of error and uncertainty; determine the type and amount of qualitative and/or quantitative data to be generated or collated

Develop

MATERIALS

- glue
- scissors

INSTRUCTIONS

Read the descriptions in Tables 5.11 and 5.12 and then cut them out and paste them into the relevant boxes in Table 5.13.

Table 5.11 Descriptions of research techniques and devices used for Parkinson's disease

MRI	PET	Animal studies	DTI
A technique that allows researchers to test the effectiveness and safety of medicines and treatments that target a specific neurological disease before they are tried on humans.	A functional neuroimaging device that enables us to see the functioning of the brain at work and highlights levels of activity (or inactivity) in various brain regions.	A technique used to investigate the structure of white matter and can detect microstructural damage, such as damage to axons, not visible on standard MRI and CT images.	A structural neuroimaging device that enables us to see images of structural changes that occur in the brain caused by neurological disorders.

Table 5.12 Descriptions of how each technique increased our understanding of Parkinson's disease

A patient is injected with an imaging drug. When the drug is fully absorbed in the patient's body, changes in their brain's dopamine-producing structures can be examined; specifically, deterioration of the metabolic pathways in the substantia nigra, the area of the brain responsible for dopamine production.	Rabbits that had been given a drug that removed dopamine from their brain went into a complete physical trance and did not respond to stimuli. However, when injected with levodopa, a drug that the brain converts to dopamine, the rabbits' movements were restored.	It can be used to evaluate changes in brain tissue caused by a neurodegenerative disease, the progression of the disease, and possible treatment responses.	Using this technique, researchers can detect people who have early-stage Parkinson's disease with 85% accuracy. It involves analysing brain networks in the basal ganglia. Using this technique, researchers have found that Parkinson's patients have much lower connectivity in the basal ganglia.

Table 5.13 Research techniques and their contributions to understanding Parkinson's disease

Description of research technique				
How the technique increased our understanding of Parkinson's disease				

5.3.3 Media analysis

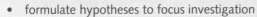

Key science skills
Develop aims and questions, formulate hypotheses and make predictions
- identify independent, dependent, and controlled variables in controlled experiments
- formulate hypotheses to focus investigation

Develop

Use the information in the article and your knowledge to answer the questions that follow.

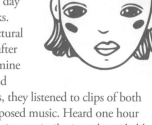

Read the article carefully and underline or highlight the key information.

Listening to favourite music improves brain plasticity

University of Toronto, *ScienceDaily*
November 9, 2021

Researchers at the University of Toronto have demonstrated that repeated listening to personally meaningful music induces beneficial brain plasticity in patients with mild cognitive impairment or early Alzheimer's disease.

Changes in the brain's neural pathways correlated with increased memory performance on neuropsychological tests, supporting the clinical potential of personalized, music-based interventions for people with dementia.

'We have new brain-based evidence that music that holds special meaning for a person, like the song they danced to at their wedding, stimulates neural connectivity in ways that help maintain higher levels of functioning,' says Dr Michael Thaut, senior author of the study.

'Typically, it's very difficult to show positive brain changes in Alzheimer's patients. These preliminary yet encouraging results show improvement in the integrity of the brain, opening the door to further research on therapeutic applications of music for people with dementia – musicians and non-musicians alike.'

The research team reported structural and functional changes in neural pathways of study participants, notably in the prefrontal cortex. Researchers showed that exposing the brains of patients with early-stage cognitive decline to autobiographical music activated a distinct neural network – a musical network – comprised of diverse brain regions that showed differences in activation after a period of daily music listening. Differences were also observed in the brain's connections and white matter, providing further evidence of neuroplasticity.

'Music-based interventions may be a feasible, cost-effective and readily accessible intervention for those in early-stage cognitive decline,' says Dr. Corinne Fischer, lead author.

For the study, 14 participants – eight non-musicians and six musicians – listened to a playlist of autobiographically relevant, long-known music for one hour a day over the course of three weeks. Participants underwent structural and fMRI scans before and after the listening period to determine changes to brain function and structure. During these scans, they listened to clips of both long-known and newly composed music. Heard one hour before scanning, the new music was similar in style yet held no personal meaning.

When participants listened to the recently heard, newly composed music, brain activity occurred mainly in the auditory cortex, centered on the listening experience. However, when participants listened to long-known music, there was significant activation in the deep-encoded network of the prefrontal cortex, a clear indication of executive cognitive engagement. There was also strong engagement in subcortical brain regions, older areas minimally affected by Alzheimer's disease pathology.

Repeated exposure to music with autobiographical meaning improved cognition in all participants, regardless of musicianship.

'Whether you're a lifelong musician or have never even played an instrument, music is an access key to your memory, your prefrontal cortex,' says Thaut.

'It's simple, keep listening to the music that you've loved all your life. Your all-time favourite songs, those pieces that are especially meaningful to you – make that your brain gym.'

Source: 'Listening to favourite music improves brain function in Alzheimer's patients: U of T research' by Josslyn Johnstone, U of T News, University of Toronto, November 10, 2021, https://www.utoronto.ca/news/listening-favourite-music-improves-brain-function-alzheimer-s-patients-u-t-research. Adapted with permission.

Questions

1 Identify the aim of this research.

_____

_____

2 Write a possible hypothesis for this study.

_____

_____

_____

3 Identify the independent variable.

_____

4 Identify one dependent variable.

5 What instrument was used to measure the dependent variable?

6 Explain why readings were taken prior to the investigation.

7 The sample included both musicians and non-musicians. Account for this feature of the design.

5.4   Chronic traumatic encephalopathy

Key knowledge
- chronic traumatic encephalopathy (CTE) as an example of emerging research into progressive and fatal brain disease

5.4.1 Understanding brain injury

Key science skills
Analyse, evaluate and communicate scientific ideas
- discuss relevant psychological information, ideas, concepts, theories and models and the connections between them

Develop

PART A

1 Using the terms in Table 5.14, label Figure 5.8.

Table 5.14 Neuron structures

Dendrite	Soma (cell body)
Microtubules (a type of cytoskeleton)	Axon
Cytoplasm	Cell membrane
Myelin	

Figure 5.8 The cytoskeleton of a healthy neuron

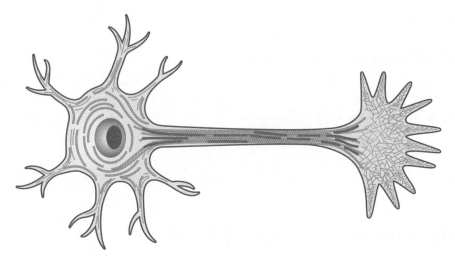

2   Draw on the diagram where the tau protein would be in a healthy neuron.

3   What is the role of the tau protein?

PART B

Answer the following questions.

1   A concussion can occur when someone receives a jolt to the skull. What happens to neurons in the brain when someone gets a concussion?

2   What are some common symptoms of concussion?

3   What is the standard treatment for a single concussion?

4   How long does it usually take for a single concussion to fully heal?

5   What chronic condition can arise from accumulating multiple head injuries? Provide both the full name (careful of your spelling!) and the commonly used acronym.

6   What is the relationship between tau proteins and brain damage?

5.4.2 Understanding the anatomy of CTE

Key science skills

Analyse, evaluate and communicate scientific ideas
- use appropriate psychological terminology, representations and conventions, including standard abbreviations, graphing conventions and units of measurement
- discuss relevant psychological information, ideas, concepts, theories and models, and the connections between them

Develop

Figure 5.9 shows two brain sections: one from a healthy, normal brain (left) and one from the brain of former University of Texas football player Greg Ploetz.

Figure 5.9 A comparison between a healthy brain (left) and a brain with CTE (right)

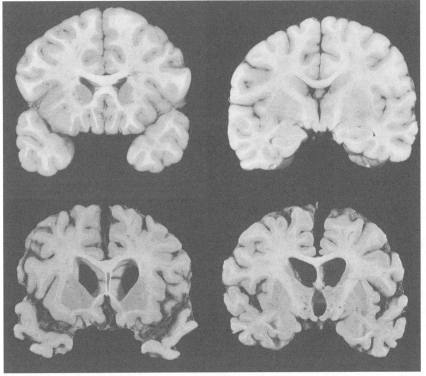

Reproduced with permission from Professor Ann McKee

Figure 5.10 provides you with the anatomical language that you will need to answer the question that follows.

Figure 5.10 Anatomy terms assist in describing the brain

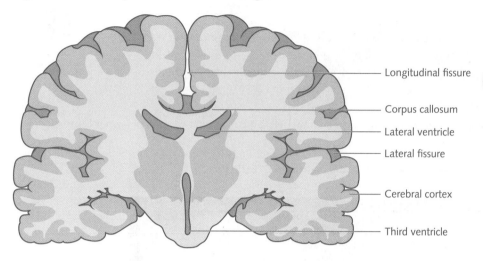

Longitudinal fissure

Corpus callosum

Lateral ventricle

Lateral fissure

Cerebral cortex

Third ventricle

This activity is seeing if you can interpret a diagram and use the information from that diagram to answer a question.

Describe three significant changes to the brain shown on the right in Figure 5.9 using the terms from Figure 5.10 of a coronal/frontal section of the brain.

_____

_____

_____

_____

_____

_____

5.4.3 Analysing data

Key science skills

Generate, collate and record data
- organise and present data in useful and meaningful ways, including tables, bar charts and line graphs

Analyse and evaluate data and investigation methods
- identify and analyse experimental data qualitatively, applying where appropriate concepts of: accuracy, precision, repeatability, reproducibility and validity; errors; and certainty in data, including effects of sample size on the quality of data obtained

Develop

PART A

A 2017 study examined the brains of 202 male players of American football and found that 177 of them had CTE, including 110 of 111 former NFL players. All the brains studied had been voluntarily donated to a 'brain bank' managed by several research institutions. Table 5.15 provides data about damage to specific regions of brains, organised by the stage of CTE at time of death.

Examine Table 5.15 and answer the questions that follow.

Table 5.15 Damage to specific brain regions as measure by tau pathology

CTE stage	No. of donors	Brain region						
		Frontal	Temporal	Parietal	Amygdala	Hippocampus	Thalamus	Cerebellum
1	11	1.1	0.6	0.2	0.4	0.1	0.3	0
2	33	1.6	1.4	1.3	1.1	0.7	0.9	0.2
3	76	2.2	2.1	1.6	2.3	2.1	1.4	0.3
4	57	2.8	2.7	2.6	2.8	2.4	2.2	0.6

Mean phosphorylated tau pathology

0			1			2			3

Note: Mean phosphorylated tau pathology is a measure of the average relative build-up of tau proteins in each brain region.

Selected data from: Mez J, Daneshavar DH, Keiman PT et al, 2017. Clinicopathological Evaluation of Chronic Traumatic Encephalopathy in Players of American Football. *Journal of the American Medical Association* (JAMA) 318(4):360–370.

1 Based on the data in Table 5.15, which two brain regions had the most advanced tau pathology?

_____

2 What types of symptoms do you think individuals with the most advanced tau pathology were most likely to experience? Justify your answer using information about brain regions and their functions. You may need to review the role of the brain areas in cognitive function.

_____

_____

_____

_____

_____

3 Which brain region consistently showed the lowest levels of tau damage (compared to other brain regions) in brains at all stages of CTE?

_____

4 What cognitive function is the least affected?

_____

PART B

Examine the data in Table 5.16 and answer the questions that follow.

Table 5.16 Symptoms of CTE

Symptoms	Found in what % of mild CTE cases (n = 27)	Found in what % of severe CTE cases (n = 84)	Found in what % of ALL CTE cases (n = 111)
Headache	30	14	18
Cognitive symptoms	**85**	**95**	**93**
Memory problems	73	92	86
Executive functioning	73	81	79
Attention	69	81	78
Language	39	66	59
Visual spatial	27	54	47
Dementia	33	85	72
Behavioral/mood symptoms	**96**	**89**	**91**
Impulsivity	89	80	82
Depressive symptoms	67	56	59
Explosivity	67	45	51
Anxiety	52	50	51
Social inappropriateness	48	32	36
Physical violence	52	28	34
Paranoia	41	31	34
Motor symptoms	**48**	**75**	**68**
Gait instability	26	66	56
Slowness	19	50	42
Coordination difficulties	26	45	41
Falling	15	46	39
Tremor	19	39	34
Difficulty swallowing	11	18	16
Slurred speech	19	13	14

Selected data from: Mez J, Daneshavar DH, Keiman PT et al, 2017. Clinicopathological Evaluation of Chronic Traumatic Encephalopathy in Players of American Football. *Journal of the American Medical Association (JAMA)* 318(4):360–370.

1   What was the most reported cognitive symptom?

_____

2   This indicates dysfunction of what part of the brain?

_____

3   What percentage of all individuals diagnosed with CTE in this study reported dementia?

_____

4   What percentage of all individuals diagnosed with CTE in this study reported behavioural/mood symptoms?

_____

5 Graph the main symptom categories of this data.

> Remember the correct way to construct a graph and to include all the elements of a well-drawn graph.

6 Considering the information about the brains used for this study, do you think the data is representative of the population? Justify your answer.

_____

_____

_____

7 What should researchers do if they want to obtain data that is more representative?

_____

_____

PART C

To test for a link between changes in midbrain structural integrity another group of researchers in 2019 correlated head hits (as measured by the helmet-worn accelerometers) with changes in white matter integrity (as measured by DTI imaging) on college footballers. Their results are shown in Figure 5.11. Answer the questions below.

Figure 5.11 The relationship between head hits and white matter integrity

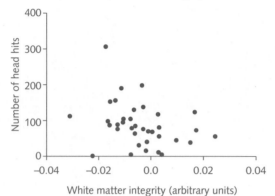

1 What is the difference between a correlational and controlled experimental methodology?

_____

_____

_____

2 Describe how DTI works.

_____

_____

_____

3 Using the data in Figure 5.11 describe the relationship between the number of hits and integrity of white matter.

_____

_____

_____

_____

5.4.4 Media analysis

Key science skills

Construct evidence-based arguments and draw conclusions
• distinguish between opinion, anecdote and evidence, and scientific and non-scientific ideas
Analyse, evaluate and communicate scientific ideas
• critically evaluate and interpret a range of scientific and media texts (including journal articles, mass media communications, opinions, policy documents and reports in the public domain), processes, claims and conclusions related to psychology by considering the quality of available evidence

Develop

Read the following article and answer the questions that follow.

'Game changer': Blood test gives Platten hope despite CTE fears

Jon Pierik, *The Age*
March 22, 2022

Four-time premiership star John Platten says the deaths of Shane Tuck and Danny Frawley have him worried about his future. This comes after West Coast premiership player and first-round draft pick, Daniel Venables, was forced into retirement last year after only 21 matches. Venables had seven brain bleeds after an incident in a marking contest against Melbourne in 2019.

Dr Rowena Mobbs, a CTE researcher and neurologist at the Macquarie University Australian Biobank, said there was major progress being made in how CTE was diagnosed, including the introduction of a blood test within five years.

'It is a game changer. The name of the game will be detecting those minute proteins that are causing the trouble in the brain using a blood test,' Dr Mobbs told The Age.

'There is a worldwide effort to help these patients with CTE. I think we will see these sorts of technologies, likewise with neuroimaging, MRI, brain scans or other types of brain scans.'

Platten, the champion rover who played in the Hawks' 1986, '88, '89 and '91 flags and is an Australian Football Hall of Fame inductee, was concussed repeatedly through his career.

'I still get a lot of headaches. I still have a bit of memory loss on a few things. I have always said I am a pretty happy-go-lucky sort of bloke but my mood swings quite a bit, especially over the last 15, 20 years,' Platten said.

'I am not saying that I have got it [CTE] but I want to make sure that I don't get it. We are all getting older … you can just feel the symptoms that are happening to me could be a part of this CTE.'

Four VFL-AFL footballers – Graham 'Polly' Farmer, Frawley, Tuck and Murray Weideman – have been posthumously diagnosed with CTE. Tuck had been suffering severe CTE when he died at the age of 38 in 2020 – his death is now before the state coroner – while the state coroner found CTE was a 'potential contributor' to the depression Frawley was experiencing before his fatal car crash in 2019.

Platten said their deaths had made him think even more about his own symptoms.

'I know both of them [Tuck and Frawley] had problems. I probably haven't got those problems as bad as those blokes. It does worry you when you hear about the CTE and what level they were at,' he said.

'I think Tucky was at level four, Danny was at three … I think the highest level is five.'

Concussion campaigner Peter Jess said there needed to be a greater focus on preventing concussions, thereby mitigating CTE, predominantly through testing using transcranial magnetic stimulation. He also wants greater punishment for 'reckless on-field behaviour' that leads to brain injuries.

'The real question is about safety,' Jess said.

The AFL continues to assess and strengthen its concussion and head-knock protocols, while the league's chief medical officer, Dr Michael Makdissi, is visiting clubs and discussing CTE.

Sporting bodies are fearful of the ramifications of CTE because of the potential financial and legal implications, and the threat of worried parents not allowing children to play sports where head knocks are more combative. But rather than fear what the new testing could find, Dr Mobbs said sports – whether that be professional or amateur – and athletes should embrace being able to make an earlier, better decision in assessing any injury.

Dr Mobbs said better testing would help all football codes, jockeys, those involved in combat sports, including boxing, those with military-related blast trauma, and help detect domestic violence.

'Some sports will be higher risk than others. We have all seen on TV the repeat head knocks, in AFL and other sports. As researchers and community members, we all love our sport and we want to see these sports thrive but in a safe way. That is so important for the children of today and we hope we don't have yet another generation facing repetitive head injuries without evidence-based care.'

Source: adapted from Pierik, J. (2022, May 22). 'Game changer': Blood test gives Platten hope despite CTE fears. *The Age*.

Read the article carefully and it might help if you underline or highlight the key information.

Questions

1   What is the current method for diagnosing CTE?

2   What protein is the article referring to as the protein responsible for the death of neurons in CTE?

3   The article mentions transcranial magnetic stimulation (TMS). What is this?

4   Shane Tuck reached level four of CTE. What symptoms would he have experienced prior to his death?

5   Identity a use of anecdote in this article.

6   Identify a use of opinion in this article.

7   Some people are of the opinion that all contact sports should be banned. Based on your understanding of CTE, what would be your response?

Chapter 5 summary

Create a mind map to show how all the ideas and concepts that you learnt about in Chapter 5 relate to each other and interlink. To create your mind map remember to complete the following steps:

1. Work out the main ideas that you want to radiate from your main idea and space these evenly around the topic.
2. Draw a line from the topic to each of your main ideas. You can write words along each line to add extra information.
3. Add another layer to your mind map to give each main idea more information.
4. Again, draw lines from the main ideas to the additional information. You can write words along each line to add even more information.

You can use colour to show more relationships within your mind map.

6 Contemporary psychological research

How do we conduct research?

Key knowledge

Scientific evidence
- the nature of evidence and information: distinction between opinion, anecdote and evidence, and between scientific and non-scientific ideas
- the quality of evidence, including uncertainty, validity and authority of data and sources of possible errors or bias
- the influence of sociocultural, economic, legal and political factors, and application of ethical understanding to science as a human endeavour
- methods of organising, analysing, and evaluating secondary data
- the use of a logbook to authenticate collated secondary data

Scientific communication
- the characteristics of effective science communication: accuracy of psychological information; clarity of explanation of scientific concepts, ideas and models; contextual clarity with reference to importance and implications of findings; conciseness and coherence; and appropriateness for purpose and audience

Analysis and evaluation of psychological research
- criteria to evaluate the validity of psychological research

6.1.1 Communicating science

Key science skills

Analyse, evaluate and communicate scientific ideas
- use clear, coherent, and concise expression to communicate to specific audiences and for specific purposes in appropriate scientific genres, including scientific reports and posters

Develop

The more accessible we can make science, the more we can spread the word.

Read the article from ScienceDaily and create a vlog transcript and storyboard in Table 6.1 that highlights the main ideas from this research.

Anxiety effectively treated with exercise

University of Gothenburg, *ScienceDaily*
9 November, 2021

Both moderate and strenuous exercise alleviate symptoms of anxiety, even when the disorder is chronic, a study led by researchers at the University of Gothenburg shows.

The study, now published in the Journal of Affective Disorders, is based on 286 patients with anxiety syndrome, recruited from primary care services in Gothenburg and the northern part of Halland County. Half of the patients had lived with anxiety for at least ten years. Their average age was 39 years, and 70 percent were women.

Through drawing of lots, participants were assigned to group exercise sessions, either moderate or strenuous, for 12 weeks. The results show that their anxiety symptoms were significantly alleviated even when the anxiety was a chronic condition, compared with a control group who received advice on physical activity according to public health recommendations.

Most individuals in the treatment groups went from a baseline level of moderate to high anxiety to a low anxiety level after the 12-week program. For those who exercised at relatively low intensity, the chance of improvement in terms of anxiety symptoms rose by a factor of 3.62. The corresponding factor for those who exercised at higher intensity was 4.88. Participants had no knowledge of the physical training or counseling people outside their own group were receiving.

'There was a significant intensity trend for improvement – that is, the more intensely they exercised, the more their anxiety symptoms improved,' states Malin Henriksson, doctoral student at Sahlgrenska Academy at the University of Gothenburg, specialist in general medicine in the Halland Region, and the study's first author.

Importance of strenuous exercise

Previous studies of physical exercise in depression have shown clear symptom improvements. However, a clear picture of how people with anxiety are affected by exercise has been lacking up to now. The present study is described as one of the largest to date.

Both treatment groups had 60-minute training sessions three times a week, under a physical therapist's guidance. The sessions included both cardio (aerobic) and strength training. A warmup was followed by circle training around 12 stations for 45 minutes, and sessions ended with cooldown and stretching.

Members of the group that exercised at a moderate level were intended to reach some 60 percent of their maximum heart rate – a degree of exertion rated as light or moderate. In the group that trained more intensively, the aim was to attain 75 percent of maximum heart rate, and this degree of exertion was perceived as high.

The levels were regularly validated using the Borg scale, an established rating scale for perceived physical exertion, and confirmed with heart rate monitors.

New, simple treatments needed

Today's standard treatments for anxiety are cognitive behavioral therapy (CBT) and psychotropic drugs. However, these drugs commonly have side effects, and patients with anxiety disorders frequently do not respond to medical treatment. Long waiting times for CBT can also worsen the prognosis.

The present study was led by Maria Åberg, associate professor at the University of Gothenburg's Sahlgrenska Academy, specialist in general medicine in Region Västra Götaland's primary healthcare organization, and corresponding author.

'Doctors in primary care need treatments that are individualized, have few side effects, and are easy to prescribe. The model involving 12 weeks of physical training, regardless of intensity, represents an effective treatment that should be made available in primary health care more often for people with anxiety issues,' Åberg says.

Source: adapted from University of Gothenburg (2021, November 9). Anxiety effectively treated with exercise. ScienceDaily. www.sciencedaily. com/releases/2021/11/211109095348.htm.

Table 6.1 Vlog transcript and associated image

Transcript	Image

6.1.2 Evaluating the integrity of sources (part 1)

Key science skills

Plan and conduct investigations
- determine appropriate investigation methodology: case study; classification and identification; controlled experiment (within subjects, between subjects, mixed design); correlational study; fieldwork; literature review; modelling; product, process or system development; simulation

Construct evidence-based arguments and draw conclusions
- use reasoning to construct scientific arguments, and to draw and justify conclusions consistent with evidence base and relevant to the question under investigation

Analyse, evaluate and communicate scientific ideas
- discuss relevant psychological information, ideas, concepts, theories and models and the connections between them
- critically evaluate and interpret a range of scientific and media texts (including journal articles, mass media communications, opinions, policy documents and reports in the public domain), processes, claims and conclusions related to psychology by considering the quality of available evidence

Develop

Order the following sources used in psychological research, from the least to the most valid and reliable, by assigning a number in the right-hand column of Table 6.2.

Table 6.2 Sources used in psychological research

Source	Rank (1–7)
Channel 7 Sunrise report	
TikTok video from a qualified scientist	
TED-Ed animation	
Opinion article from a journalist	
Peer reviewed article	
Review article from ScienceDaily	
Facebook post	

6.1.3 Evaluating the integrity of sources (part 2)

Key science skills

Analyse and evaluate data and investigation methods

- evaluate investigation methods and possible sources of error or uncertainty, and suggest improvements to increase validity and to reduce uncertainty

Construct evidence-based arguments and draw conclusions

- use reasoning to construct scientific arguments, and to draw and justify conclusions consistent with evidence base and relevant to the question under investigation
- identify, describe and explain the limitations of conclusions, including identification of further evidence required

Analyse, evaluate and communicate scientific ideas

- discuss relevant psychological information, ideas, concepts, theories and models and the connections between them
- critically evaluate and interpret a range of scientific and media texts (including journal articles, mass media communications, opinions, policy documents and reports in the public domain), processes, claims and conclusions related to psychology by considering the quality of available evidence

Develop

Read the following article and evaluate its validity and reliability using the CRAP technique. Details about the website, header, contact details have also been included. Underlined words indicate a link to another source.

Just because it is out there, doesn't mean it is trustworthy.

Why multitasking doesn't work

Studies show it makes us less efficient and more prone to errors

Cleveland Clinic
10 March, 2021

Some days you feel like a master multitasker but did you know that for most people, in most situations, multitasking isn't actually possible?

We're really wired to be monotaskers, meaning that our brains can only focus on one task at a time, says neuropsychologist Cynthia Kubu, PhD. 'When we think we're multitasking, most often we aren't really doing two things at once, but instead, we're doing individual actions in rapid succession, or task-switching,' she says.

One study found that just 2.5% of people are able to multitask effectively. For the rest of us, our attempts to do multiple activities at once aren't actually that.

Studies show that when our brain is constantly switching gears to bounce back and forth between tasks – especially when those tasks are complex and require our active attention – we become less efficient and more likely to make a mistake.

This might not be as apparent or impactful when we're doing tasks that are simple and routine, like listening to music while walking, or folding laundry while watching TV. But when the stakes are higher and the tasks are more complex, trying to multitask can negatively impact our lives – or even be dangerous.

So-called multitasking divides our attention. It makes it harder for us to give our full attention to one thing. For example, in studies, attempting to complete additional tasks during a driving simulation led to poorer driving performance. Other studies suggest that people who frequently 'media multitask' (like listening to music while checking email or scrolling through social media while watching a movie) are more distracted and less able to focus their attention even when they're performing only one task.

It can also affect our ability to learn, because in order to learn, we need to be able to focus.

'The more we multitask, the less we actually accomplish, because we slowly lose our ability to focus enough to learn,' Dr. Kubu says. 'If we're constantly attempting to multitask, we don't practice tuning out the rest of the word to engage in deeper processing and learning.' One study found that college students who tried to multitask took longer to do their homework and had lower average grades.

Another pitfall is that trying to do too much at once makes it harder to be mindful and truly present in the moment – and mindfulness comes with a plethora of benefits for our minds and our bodies. In fact, many therapies based on mindfulness can even help patients suffering from depression, anxiety, chronic pain and other conditions.

Opting to focus on one task at a time can benefit many aspects of our life, including the workplace.

Take surgeons, for example. 'People assume that a surgeon's skill is primarily in the precision and steadiness of his or her hands. While there's some truth to that, the true gift of a surgeon is the ability to single-mindedly focus on one person and complete a series of tasks over the course of many hours,' Dr. Kubu xplains.

But surgeons aren't necessarily born with this ability to monotask. Rather, they develop and perfect it through hours of practice. And you can, too.

'You don't need to be a surgeon to benefit from freeing yourself of the pressure to multitask,' Dr. Kubu says. 'Whether it's taking a long road trip, organizing an event or reading a book, we unequivocally perform best one thing at a time. I encourage you to give it a try.'

Source: adapted from Cleveland Clinic. (2021, March 10). Why multitasking doesn't work. https://health.clevelandclinic.org/science-clear-multitasking-doesnt-work/

Table 6.3 Outline of the CRAP evaluation technique

Question to ask yourself	Evaluation
Currency: the timeliness of the information	
Was the information published or updated recently?	
Is the information recent enough for your topic?	
If using a website, do the links within the article work?	
Reliability: the accuracy and truthfulness of the information	
Is the information supported by evidence like data or quotes? Are there references for the evidence?	
Does the source make reasonable claims about what the evidence shows? Has the information (or its references) been reviewed?	

Question to ask yourself	Evaluation
Has the study been replicated?	
Can you confirm the information using another source?	
Does the language or tone seem unbiased and professional?	
Authority: the source of the information	
Is the author, publisher or sponsor of the information a trustworthy source, such as an educational or government institution?	
Is the author qualified to write on the topic?	
Is the author likely to be unbiased about the topic?	
Is there any contact information?	
Purpose: the reasons the information exists	
Is the purpose of the information to teach or inform rather than to sell, entertain or persuade?	
Is the information fact, rather than opinion or anecdote? Does the source's point of view seem unbiased?	

Source: adapted from Charles Sturt University, https://libguides.csu.edu.au/EML102/CRAP-test

6.2

Ethical understandings of research and technology

Key knowledge
Scientific communication
- psychological concepts specific to the investigation: definitions of key terms; and use of appropriate psychological terminology, conventions and representations
- the influence of sociocultural, economic, legal and political factors, and application of ethical understanding to science as a human endeavour

6.2.1 Analysing ethics in research

Key science skills
Comply with safety and ethical guidelines
- demonstrate ethical conduct and apply ethical guidelines when undertaking and reporting investigations

Construct evidence-based arguments and draw conclusions
- use reasoning to construct scientific arguments, and to draw and justify conclusions consistent with evidence base and relevant to the question under investigation

Analyse, evaluate and communicate scientific ideas
- critically evaluate and interpret a range of scientific and media texts (including journal articles, mass media communications, opinions, policy documents and reports in the public domain), processes, claims and conclusions related to psychology by considering the quality of available evidence
- analyse and evaluate psychological issues using relevant ethical concepts and principles, including the influence of social, economic, legal and political factors relevant to the selected issue

Develop

PART A

1   Read the two articles that introduce a potential ethical issue in current research and treatment of the use of CBD (cannabidiol oil) in treating physical and mental health issues.

Research shows pain relieving effects of CBD

Syracuse University, *ScienceDaily*
25 April, 2021

It's been hailed as a wonder drug and it's certainly creating wonder profits. By some estimates, the Cannabidiol (or CBD) market could be worth $20 billion dollars by 2024.

While users tout its effectiveness in pain relief, up until now there's been limited experimental human research on the actual effectiveness of the drug. However, a new study led by University researchers sheds light on the ability of CBD to reduce pain along with the impact that the so-called placebo effect may have on pain outcomes.

'For science and the public at large the question remained, is the pain relief that CBD users claim to experience due to pharmacological effects or placebo effects,' says Martin De Vita, a researcher in the psychology department in the College of Arts and Sciences. 'That's a fair question because we know that simply telling someone that a substance has the ability to relieve their pain can actually cause robust changes in their pain sensitivity. These are called expectancy effects.'

De Vita, along with Stephen Maisto, research professor and professor emeritus of psychology, were uniquely prepared to answer that exact question. The pair, along with fellow lab member and doctoral candidate Dezarie Moskal, previously conducted the first systematic review and meta-analysis of experimental research examining the effects cannabinoid drugs on pain.

As the first experimental pain trial to examine CBD, their study yielded consistent and noteworthy results. Among other findings, the data showed that CBD and expectancies for receiving CBD do not appear to reduce experimental pain intensity, but do make the pain feel less unpleasant.

De Vita and Maisto used sophisticated equipment that safely induces experimental heat pain, allowing them to measure how the recipient's nervous system reacts and responds to it. 'Then we administer a drug, like pure CBD, or a placebo and then re-assess their pain responses and see how they change based on which substance was administered,' says De Vita.

Researchers then took it a step farther by manipulating the information given to participants about which substances they received. In some cases, participants were told that they got CBD when they actually received a placebo, or told they would be getting a placebo when they actually got CBD.

'That way we could parse out whether it was the drug that relieved the pain, or whether it was the expectation that they had received the drug that reduced their pain,' according to De Vita. 'We hypothesized that we would primarily detect expectancy-induced placebo analgesia (pain relief). What we found though after measuring several different pain outcomes is that it's actually a little bit of both. That is, we found improvements in pain measures caused by the pharmacological effects of CBD and the psychological effects of just expecting that they had gotten CBD. It was pretty remarkable and surprising.'

'The data is exciting but pretty complex in that different pain measures responded differently to the drug effect, to the expectancy, or both the drug and expectancy combined – so we're still trying to figure out what is behind the differential data with different kinds of pain measures,' said Maisto. 'The next step is studying the mechanisms underlying these findings and figuring out why giving instructions or CBD itself causes certain reactions to a pain stimulus.'

Most people think of pain as an on and off switch, you either have it or you don't. But pain, as De Vita describes it, is a complex phenomenon with several dimensions influenced by psychological and biological factors.

For example, whereas pain intensity reflects a 'sensory' dimension of pain, unpleasantness represents an 'affective,' or emotional, aspect of pain. 'If you think of pain as the noxious noise coming from a radio the volume can represent the intensity of the pain, while the station can represent the quality,' says De Vita.

Results from his previous study showed that while cannabinoid drugs weren't reducing the volume of pain, they were 'changing the channel making it a little less unpleasant.' According to De Vita, 'It's not sunshine and rainbows pleasant, but something slightly less bothersome. We replicated that in this study and found that CBD and expectancies didn't significantly reduce the volume of the pain, but they did make it less unpleasant – it didn't bother them as much.'

As part of the study De Vita and Maisto developed advanced experimental pain measurement protocols 'to pop the hood and start looking at some of these other mechanistic pain processes,' says De Vita. 'It's not just pain, yes or no, but there are these other dimensions of pain, and it would be interesting to see which ones are being targeted. We found that sometimes pharmacological effects of CBD brought down some of those, but the expectancies did not. Sometimes they both did it. Sometimes it was just the expectancy. And so, we were going into this thinking we were going to primarily detect the expectancy-induced pain relief but what we found out was way more complex than that and that's exciting.'

One important note to also consider is the source of the CBD. 'What we used in our study was pure CBD isolate oil,' says De Vita. 'Commercially available CBD products differ in their content and purity, so results might be different for different CBD products, depending on what other compounds they may or may not contain.'

Source: adapted from Syracuse University (2021, April 25). Research shows pain relieving effects of CBD. *ScienceDaily*. www.sciencedaily.com/releases/2021/04/210423130221.htm.

CBD oil benefits vs. side effects

While it may be helpful, it may not be safe for all

Cathy Wong

Medically reviewed by Meredith Bull, ND
28 June, 2022

What exactly is CBD oil?

CBD oil is a hemp plant extract known as cannabidiol mixed with a base (carrier) oil like coconut oil or hemp seed oil. CBD oil comes from *Cannabis indica* and *Cannabis sativa* plants.

CBD oil benefits

People who support the use of CBD claim that CBD oil benefits people with a variety of health problems. CBD oil is said to be good for: Anxiety, Cancer, Chronic pain, Depression Insomnia, Parkinson's disease.

As CBD has gained popularity, researchers have been trying to study it more. Still, there has not yet been a lot of clinical research focused on finding evidence to back up these health claims.

Here's a deeper dive into what is known about a few of the purported health benefits of CBD oil.

Anxiety

A 2015 review of research that was published in the journal Neurotherapeutics suggested that CBD might help treat anxiety disorders.

The study authors reported that CBD had previously shown powerful anxiety-relieving effects in animal research—and the results were kind of surprising.

In most of the studies, lower doses of CBD (10 milligrams per kilogram, mg/kg, or less) improved some symptoms of anxiety, while higher doses (100 mg/kg or more) had almost no effect.

The way that CBD acts in the brain could explain why this happens. In low doses, CBD might act the same as the surrounding molecules that normally bind to the receptor that 'turns up' their signalling. However, at higher doses, too much activity at this receptor site could produce the opposite effect.

There have not been many trials to look at CBD's anxiety-relieving effects in humans. However, one was a 2019 study published in the Brazilian Journal of Psychiatry.

For the study, 57 men took either CBD oil or a sugar pill with no CBD in it (placebo) before a public-speaking event.

The researchers assessed the participants' anxiety levels using measures like blood pressure and heart rate. The researchers also used a reliable test for mood states called the Visual Analog Mood Scale (VAMS).

The men who took 300 milligrams (mg) of CBD oil reported less anxiety than the men who were given a placebo; however, the men who took 100 mg or 600 mg of CBD oil did not experience the same effects.

Cancer

Proponents say CBD oil has benefits for people with cancer. Although some studies have shown promise, there have been no large studies proving the benefits of CBD oil as a cancer treatment.

Other studies suggest that CBD might interact with cancer drugs.

If you have cancer and are considering CBD, talk to your oncologist first about whether or not it is safe for you to use.

Sleep

Proponents say CBD oil has benefits as a sleep aid, but research so far is inconclusive.

A 2017 review pointed out that many studies have been small and limited. However, the authors also noted that because cannabinoids seem to have an effect on the sleep-wake cycle, their potential as a sleep aid is worthy of additional research.

Adapted from: Wong, C. (2022, June 28). CBD oil benefits vs. side effects. https://www.verywellhealth.com/cbd-oil-benefits-uses-side-effects-4174562.

2 Complete an analysis of the use of cannabidiol oil in the treatment of mental and physical illnesses.

Table 6.4 Ethical analysis

Technology or treatment	Cannabidoil as a treatment for mental and physical illnesses	
Brainstorm of initial thoughts and feelings		

Guiding principle	Description	Analysis
Integrity	Honesty and transparency	
Respect	Every individual has the right to autonomy and to choose their own course of action	
Beneficence	A duty to do good	
Non-maleficence	The duty to minimise harm	
Justice	Equal distribution of benefits, risks, costs and resources	

PART B

Determine possible implications by evaluating the issue based on the factors listed in Table 6.5.

Table 6.5 Ethical factors

Factors	Evaluation
Political	
Social	
Economic	
Legal	

PART C

1   Construct two paragraphs that summarise the information you have outlined in Table 6.6 using the TEEL structure.

Table 6.6 Constructing paragraphs

Paragraph 1	
T	
E	

»

»

E	
L	

Paragraph 2	
T	
E	
E	
L	

2 Write your response here:

_____

_____

_____

_____

_____

_____

_____

_____

Social cognition

7

7.1 **Making sense of the social world**

Key knowledge
- the role of person perception, attributions, attitudes, and stereotypes in interpreting, analysing, remembering, and using information about the social world, including decision-making and interpersonal interactions

7.1.1 Person perception

Key science skills
Analyse and evaluate data and investigation methods
- process quantitative data using appropriate mathematical relationships and units, including calculations of percentages, percentage change and measures of central tendencies (mean, median, mode), and demonstrate an understanding of standard deviation as a measure of variability
Generate, collate and record data
- systematically generate and record primary data, and collate secondary data, appropriate to the investigation

Develop

1 Match each person in Table 7.1 to their occupation in Table 7.2 by writing the corresponding number in the first column.

Table 7.1 Images of people from the Internet

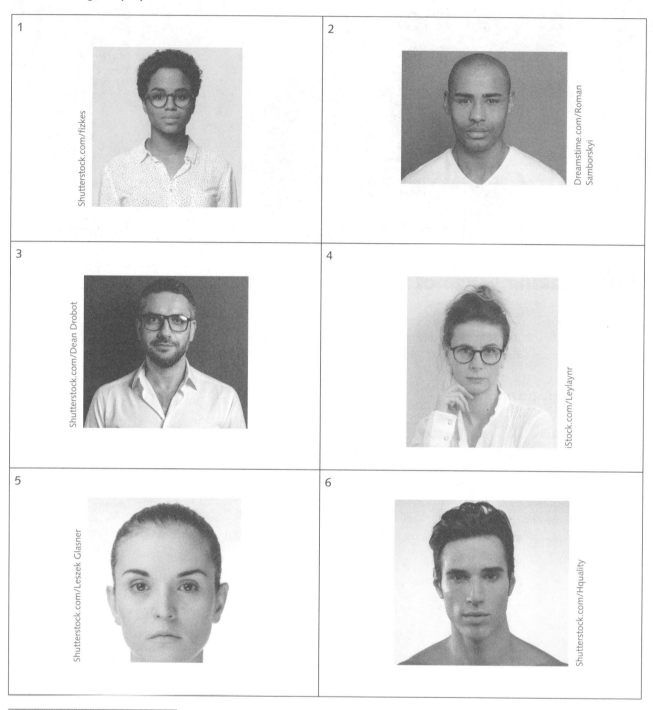

Table 7.2 Matching occupations

Person	Occupation
	Model
	Accountant
	Computer programmer
	Tradesperson
	Prison warden
	Neuroscientist

2 Collate a set of class results in Table 7.3 by recording the frequency of each person the class placed under each occupation.

Table 7.3 Class data

Person	Occupation	Highest frequency	Percentage (%)
	Model		
	Accountant		
	Computer programmer		
	Tradesperson		
	Prison warden		
	Neuroscientist		

3 Convert the highest frequency (tally) per occupation into a percentage.

_____

4 Describe any patterns in your results.

_____

_____

We all use different criteria when making decisions.

5 What criteria did you use to help you decide which person belonged to each occupation?

_____

_____

_____

6 List two criteria per person and write these in Table 7.4.

Table 7.4 Criteria used to decide occupation

Person	Criterion
1	
2	
3	
4	
5	
6	

7 Categorise each criterion using the three key components of person perception outlined in the textbook. Write these in Table 7.5.

Table 7.5 Using components of person perception to categorise people

Component	Criterion
Physical cues	
Saliency detection	
Social categorisation	

8 Share your criteria with your class. Describe any patterns.

7.1.2 Attributions

Key science skills

Analyse, evaluate and communicate scientific ideas
- discuss relevant psychological information, ideas, concepts, theories and models and the connections between them

Develop

Read the following story and answer the questions that follow.

Once upon a time, a husband and wife lived together in a part of the city separated by a river from the places of employment, shopping and entertainment. The husband had to work nights. Each evening he left his wife and took the ferry to work, returning in the morning.

The wife soon tired of this arrangement. Restless and lonely, she would take the next ferry into town each evening and develop relationships with a series of lovers. Anxious to preserve her marriage, she always returned home before her husband. In fact, her relationships were always limited. When they threatened to become too intense, she would precipitate a quarrel with her current lover and begin a new relationship.

One night she caused such a quarrel with a man we will call Lover 1. He slammed the door in her face, and she started back to the ferry. Suddenly she realized that she had forgotten to bring the money for her return fare. She swallowed her pride and returned to Lover 1's apartment to borrow the fare. After all, she did have to get home. But Lover 1 was vindictive and angry because of the quarrel. He slammed the door on his former lover, leaving her with no money. She remembered that a previous lover, who we shall call Lover 2, lived just a few doors away.

Surely, he would give her the ferry fare. However, Lover 2 was still so hurt from their old quarrel that he, too, refused her the money.

Now the hour was late, and the woman was getting desperate. She rushed down to the ferry and pleaded with the captain. He knew her as a regular customer. She asked if he could let her ride free and if she could pay the next night. But the captain insisted that rules were rules, and that he could not let her ride without paying the fare.

Dawn would soon be breaking, and her husband would be returning from work. The woman remembered that there was a free bridge about a mile further on. But the road to the bridge was a dangerous one. Nonetheless, she had to get home, so she took the road. On the way, a random man stepped of the bushes and demanded her money. She told him she had none. He seized her. In the ensuing tussle, the man stabbed the woman, and she died.

Thus ends our story. There have been six characters: husband, wife, Lover 1, Lover 2, the captain, and random man.

Sources: Dolgoff, R., & Feldstein, D. (1984). Understanding social welfare (2nd ed.). Longman; and Myers, D. G. (1993). Social psychology (4th ed.). McGraw-Hill.

1 List in descending order of responsibility for this woman's death, all the characters. In other words, the one most responsible is listed first; the next most responsible, second; and so forth. You may want to do this activity with a larger sample of participants.

Table 7.6 Who is responsible?

Responsibility (most to least)	Character
1	
2	
3	
4	
5	
6	

Scan through the story again if you cannot remember.

2 Explain your choices using the following terms:

a. Internal attribution

_____

_____

_____

b. External attribution

_____

_____

_____

c. Fundamental attribution error

_____

_____

_____

3 Would the responsibility of the woman's death change if you were told that she was a widow and needed to cross the river to earn extra money to help with raising her family? Explain.

_____

_____

_____

7.1.3 Attitude formation

Key science skills

Analyse, evaluate and communicate scientific ideas

- discuss relevant psychological information, ideas, concepts, theories and models and the connections between them
- analyse and explain how models and theories are used to organise and understand observed phenomena and concepts related to psychology, identifying limitations of selected models/theories

Develop

PART A

Study the terms in Table 7.7, then use them to fill in the blanks in the following paragraphs.

Table 7.7 Terms associated with attitude formation

learned	formed	experiences
interaction	manipulates	media
evaluations	environment	practices
association	contact	common

Attitudes are _____ we make about ourselves, others, objects
and _____. They are _____ through exposure to
the _____ and can be either positive or negative.

Direct _____ is one way that attitudes can be _____ .
This includes personal experiences and _____ with others.

People also form a particular attitude as a result of child rearing, which includes parental values, beliefs
and _____.

Attitudes can also be formed through group membership due to _____ with people
with whom we share _____ characteristics.

The _____ can also have a powerful impact on the way attitudes are formed and
often _____ the audience through the use of the Internet, television, radio and magazines.

PART B

1   Study the information in Table 7.8.
2   Study the incomplete diagram of the tri-component model of attitudes
    (Figure 7.1).
3   Choose the correct definitions from Table 7.8 and write them in the correct
    spaces in Figure 7.1 to define 'Attitude', and the 'Cognitive', 'Affective' and
    'Behavioural' components.
4   Provide your own definition of the tri-component model of attitudes and
    examples of the cognitive and behavioural components in the 'Example' boxes.

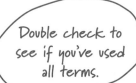

Double check to see if you've used all terms.

Table 7.8 Definitions of terms related to attitudes

Refers to how a person feels emotionally about various people, objects, institutions, events or issues	A learned idea about ourselves and others, objects, institutions, events or issues
Refers to what a person thinks about various people, objects, institutions, events or issues, and the reasons why they think this	Refers to how a person acts towards various people, objects, institutions, events or issues

Figure 7.1 The tri-component model of attitudes

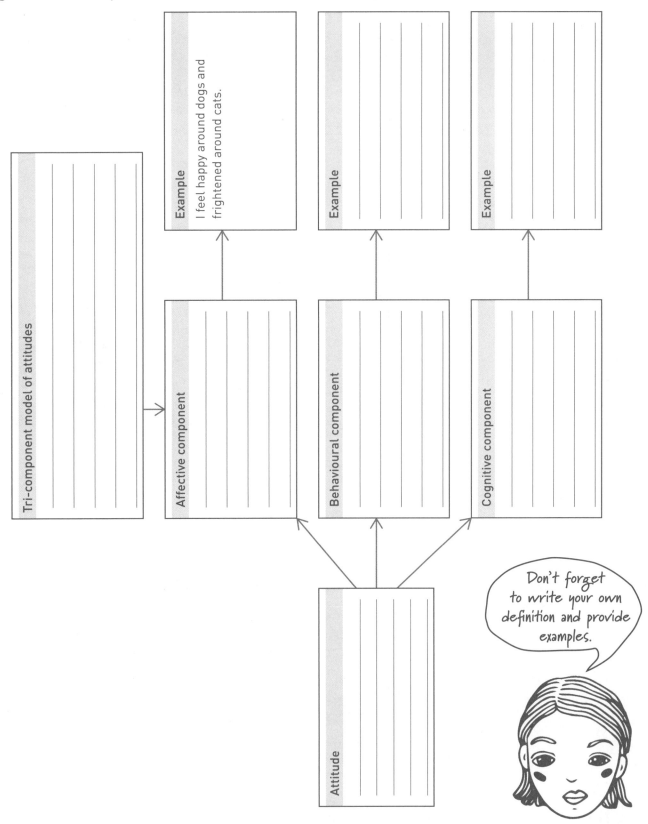

7.1.4 Components of attitude

Key science skills
Analyse, evaluate and communicate scientific ideas
- discuss relevant psychological information, ideas, concepts, theories and models and the connections between them
- analyse and explain how models and theories are used to organise and understand observed phenomena and concepts related to psychology, identifying limitations of selected models/theories

Develop

PART A

In the second column of Table 7.9, write whether each activity in the first column involves the affective, behavioural or cognitive component of an attitude.

Use the words in column one to help you determine the attitude component.

Table 7.9 Components of attitude

Activity	Attitude component (affective, behavioural, cognitive)
Knitting	
Imagining	
Believing	
Hurting	
Being happy	
Daydreaming	
Eating	
Being tired	
Silently making a wish	
Doubting	
Running slowly	
Being angry	
Playing the trumpet	
Preferring dark chocolate to milk chocolate	
Sleeping	

PART B

Consider the following attitude statements and identify the affective, cognitive and behavioural components for each.

1   Public transport users should not be permitted to smoke at bus, tram and train terminals because of the effects of passive smoking on other passengers. Besides, I dislike smokers.
- Affective component: _____
- Cognitive component: _____
- Behavioural component: _____

2 People who don't dress appropriately for job interviews don't deserve to get the job. I would not employ someone unless they had made an effort with their appearance.

   - Affective component: _____
   - Cognitive component: _____
   - Behavioural component: _____

3 I would rather watch football on television than go to the game, because I hate crowds and it is too difficult to see the play on the opposite side of the field. The game is more comfortably viewed while I am sitting on the couch at home.

   - Affective component: _____
   - Cognitive component: _____
   - Behavioural component: _____

4 There should be harsher penalties for people who drink and drive. My mother was hit by a drunk driver while she was out walking the dog. I will never drink and drive when I get my licence.

   - Affective component: _____
   - Cognitive component: _____
   - Behavioural component: _____

5 All of my friends are getting facial piercings; they look really good, and I would like to get one. My parents won't let me because they say they can cause damage and leave a scar.

   - Affective component: _____
   - Cognitive component: _____
   - Behavioural component: _____

6 I don't think P-plate drivers should have passenger restrictions because it makes it harder to organise a designated driver when they go out; and if they can carry passengers, it may mean less drink driving. I would vote against this law if given the chance.

   - Affective component: _____
   - Cognitive component: _____
   - Behavioural component: _____

7 People who shoot ducks for fun should think about taking up a new sport. It is cruel to the animals to shoot them, and ducks should be protected. I would never go duck shooting.

   - Affective component: _____
   - Cognitive component: _____
   - Behavioural component: _____

8 Skydiving is the best fun in the world. I think that everyone should have a go at it, because it makes you feel terrific. I am going to save up my money to go again.

   - Affective component: _____
   - Cognitive component: _____
   - Behavioural component: _____

7.1.5 Role of stereotyping in attitude formation

Key science skills
Analyse, evaluate and communicate scientific ideas
- discuss relevant psychological information, ideas, concepts, theories and models and the connections between them
- analyse and explain how models and theories are used to organise and understand observed phenomena and concepts related to psychology, identifying limitations of selected models/ theories

Develop

PART A

We not only stereotype people, but we can also stereotype postcodes and countries.

On the map in Figure 7.2, come up with some of the stereotyped descriptions that are often expressed in the media for at least three world areas (continent or country). A common stereotype in Australia has been marked up for you.

Figure 7.2 World map

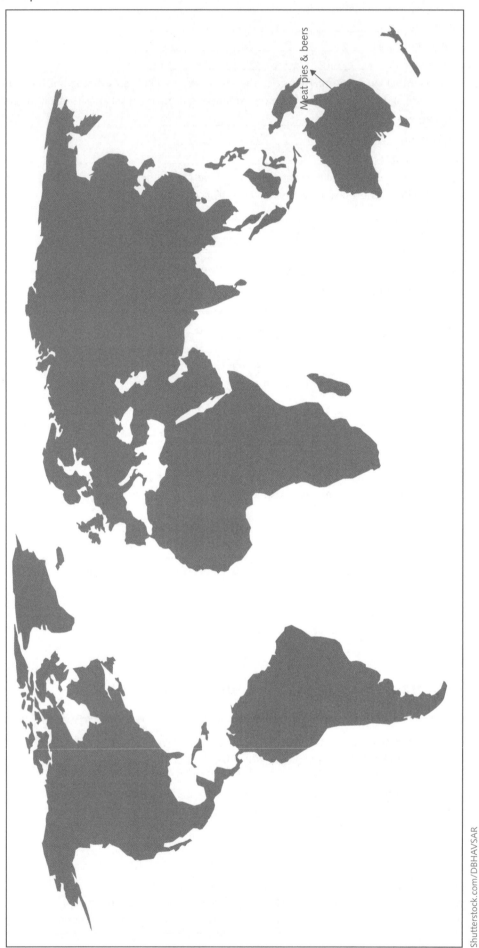

PART B

Apply the tri-component model of attitudes to each area.

1 Area _____

   • Affective component: _____

   • Cognitive component: _____

   • Behavioural component: _____

2 Area _____

   • Affective component: _____

   • Cognitive component: _____

   • Behavioural component: _____

3 Area _____

   • Affective component: _____

   • Cognitive component: _____

   • Behavioural component: _____

7.2 Cognitive dissonance and cognitive bias

Key knowledge
• the avoidance of cognitive dissonance using cognitive biases

7.2.1 Cognitive dissonance

Key science skills
Analyse, evaluate and communicate scientific ideas
• discuss relevant psychological information, ideas, concepts, theories and models and the connections between them
• analyse and explain how models and theories are used to organise and understand observed phenomena and concepts related to psychology, identifying limitations of selected models/theories
• critically evaluate and interpret a range of scientific and media texts (including journal articles, mass media communications, opinions, policy documents and reports in the public domain), processes, claims and conclusions related to psychology by considering the quality of available evidence

Develop

Table 7.10 describes one of the limitations to the tri-component model of attitudes, which often leads to cognitive dissonance.

Use the information in the table to answer the following questions.

1 What is this limitation?

_____

2 In the column on the right, describe how each person in the cartoon would reduce any possible cognitive dissonance that may result from this limitation.

Table 7.10 Reducing cognitive dissonance

Cartoon	Reducing cognitive dissonance
HOW GROSS! GET THAT DEAD THING OUT OF MY KITCHEN! Cartoon © Mark Parisi, www.offthemark.com	
HEY, IT'S BAD FOR YOUR EYES TO SIT SO CLOSE TO THE SCREEN… offthemark.com Cartoon © Mark Parisi, www.offthemark.com	

7.2.2 Cognitive bias

Key science skills

Analyse, evaluate and communicate scientific ideas

- discuss relevant psychological information, ideas, concepts, theories and models and the connections between them

Develop

Use the terms in Table 7.11 to identify the type of cognitive bias demonstrated in each scenario.

Table 7.11 Types of cognitive bias

self-serving bias	hindsight bias	misinformation effect
actor-observer bias	halo effect	false-consensus bias
anchoring bias	the Dunning-Kruger effect	optimism bias
attentional bias	functional fixedness	confirmation bias

1   You have texted your friend to meet up after school. You arrive on time, but your friend is 30 minutes late. You are not happy; you think that your friend is late because they have no regard for you or your time.

2   You have recently purchased a couple of plain white t-shirts from Tommy Hilfiger for $50 each. The next day you see similar t-shirts from Target for $10. You assume that the quality of the Target t-shirts is poor.

3   You are trying to keep to healthy eating and good sleep hygiene habits. One Friday night after completing two SACs in two days, you meet your friends at the food hall for dinner. You are very hungry and tired. You know you should order a chicken salad, but you smell the cheese pizza so order that instead.

4   Your friend is sure that their maths teacher does not like them. You haven't picked up on this, so you decide to observe more closely. At the end of the lesson your friend comes up to you and says, 'Did you see how Mrs Jones picked on me? She asked me to answer questions she knew I didn't know! I told you she didn't like me.' While you did see that Mrs Jones asked your friend to answer a question, you also saw Mrs Jones smile at your friend and ask other people in the class to answer questions.

5   You and your friends are strong advocates for alternative sources of power to coal-fired power plants. You voted for the Greens at the last Federal election. You assume your friends did as well.

6   You spend the afternoon in the library researching for a history assignment. You are part way through, and the Internet goes down. You leave the library with your task unfinished.

7   Your last pair of kicks were from Nike, and they were great. You purchase all your sports clothes from Nike as they are the best brand.

8   The Bulldogs have been sitting at around 6th on the AFL ladder. They are playing against a team that is 4th. The Bulldogs win by 4 points. You tell everyone that you knew that they would win.

9   You are writing a dream diary for school. In this diary you are required to write down every detail. Later that week you recount one of your dreams to the class, adding information when you are asked for more detail. When you re-read your diary entry there are about six pieces of information that are not included in the entry.

10   Jane is a bright, enthusiastic chef keen to start her own restaurant. She opens a restaurant in a building where six other businesses failed within the first three years. Jane, however, felt that she had what it took to make her restaurant succeed.

11   Your netball team won three games in a row. You are proud of the hard work and effort you have been putting into your training. Last week your team lost. You think the reason why is that the referees were too young and inexperienced.

12   One of your friends has this annoying habit of telling jokes at parties. They think they are hilarious. They are not.

7.3 Heuristics

Key knowledge
- the positive and negative influences of heuristics as mechanisms for decision making and problem-solving

7.3.1 Applying heuristics

Key science skills
Analyse, evaluate and communicate scientific ideas
- discuss relevant psychological information, ideas, concepts, theories and models and the connections between them

Develop

Advertisers rely on heuristics to influence consumer decision-making.

1 For each pair of products in Table 7.12, *quickly* decide which product you will purchase and then apply how heuristics influenced your decision.

Table 7.12 Consumer products

Brand: Mars	Brand: Snickers
Which chocolate are you likely to purchase?	
Which heuristic did you use? Justify	

Brand: Theory

Brand: Nike

Which brand of t-shirt would you most likely purchase?

Which heuristic did you use? Justify.

Heuristics can be helpful when we need to make decisions.

Brand: Toyota

Brand: Toyota

You only have the choice of the two vehicles shown, which one are you likely to purchase?

Which heuristic did you use? Justify.

7.4   The effects of prejudice and discrimination

Key knowledge
- the influence of prejudice and discrimination within society on a person's and/or group's mental wellbeing and ways to reduce it

7.4.1 Prejudice and discrimination

Key science skills
Analyse, evaluate and communicate scientific ideas
- discuss relevant psychological information, ideas, concepts, theories and models and the connections between them

Develop

PART A

Study the terms in Table 7.13, then use them to fill in the blanks in the statements that follow.

Table 7.13 Terms associated with prejudice

sexism	stereotype	discrimination	ageism
affective	behavioural	prejudice	racism

1 _____ refers to an emotional attitude held towards members of a specific social group.

2 An oversimplified image of people who belong to a particular group that exaggerates group characteristics and ignores individual strengths and weaknesses is a(n) _____.

3 When prejudice leads you to exclude or reject a person based on their gender, you are demonstrating _____.

4 _____ refers to the unequal treatment of people who should have the same rights as others.

5 If an employer rejected an applicant for a job on the grounds that they were too young, the employer has demonstrated _____.

6 According to the tri-component model of attitudes, prejudice is an example of a(n) _____ attitude component.

7 If a nightclub refused admittance to people based on the colour of their skin, this would be an example of _____.

8 According to the tri-component model of attitudes, discrimination is an example of a(n) _____ attitude component.

PART B

1   Study this list of dot points describing prejudice and discrimination:

- Treating someone unfairly or differently because they belong to (or are perceived to belong to) a particular group of people
- An attitude that is usually supported by little or no direct evidence
- A response or action that results in the unequal treatment of people who should have the same rights as others
- Can be positive, if it provides special advantages to an individual or group
- Can be negative, if it limits opportunities to an individual or group
- A negative emotional attitude towards thinking about or interacting with members of a specific social group
- Originates from suspicion, fear or hatred.

2   Complete the Venn diagram of the similarities and differences between prejudice and discrimination (Figure 7.3) by writing each dot point under either the 'Prejudice' heading or the 'Discrimination' heading.

Figure 7.3 A Venn diagram of similarities and differences between prejudice and discrimination

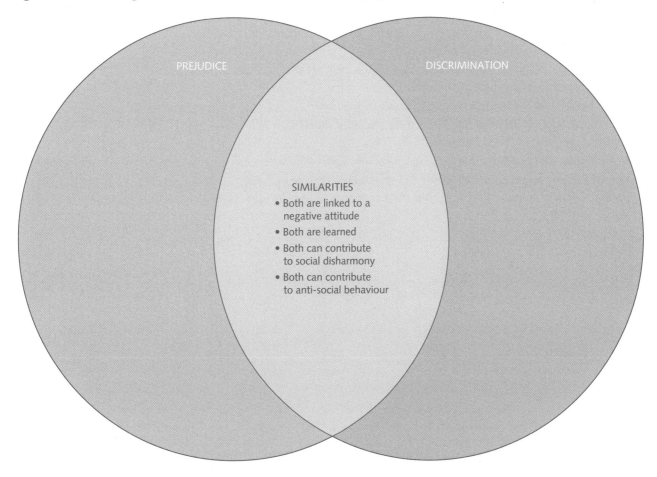

PREJUDICE

DISCRIMINATION

SIMILARITIES
- Both are linked to a negative attitude
- Both are learned
- Both can contribute to social disharmony
- Both can contribute to anti-social behaviour

7.4.2 Factors that contribute to, or reduce, prejudice

Key science skills

Analyse, evaluate and communicate scientific ideas

- discuss relevant psychological information, ideas, concepts, theories and models and the connections between them

Develop

You'll need some extra materials for this one.

MATERIALS

- A3 poster paper
- scissors
- glue

INSTRUCTIONS

1. Study Figure 7.4. In the triangle, write a definition of prejudice.
2. In the boxes below the triangle, identify and explain four factors that could contribute to the development of prejudice.
3. In the 'Example' rectangles, give brief examples of how each of the factors can contribute to the development of prejudice.
4. Study Figure 7.5. In the triangle, write a definition of prejudice.
5. In the boxes above the triangle, identify and explain four factors that could contribute to the reduction of prejudice.
6. In the 'Example' rectangles, give a brief example of how each of the factors can contribute to the reduction of prejudice.
7. Cut out Figures 7.4 and 7.5 and glue them to your A3 sheet of paper or poster cardboard and add the title 'Factors that contribute to, or reduce, prejudice'.

Figure 7.4 Factors contributing to the development of prejudice

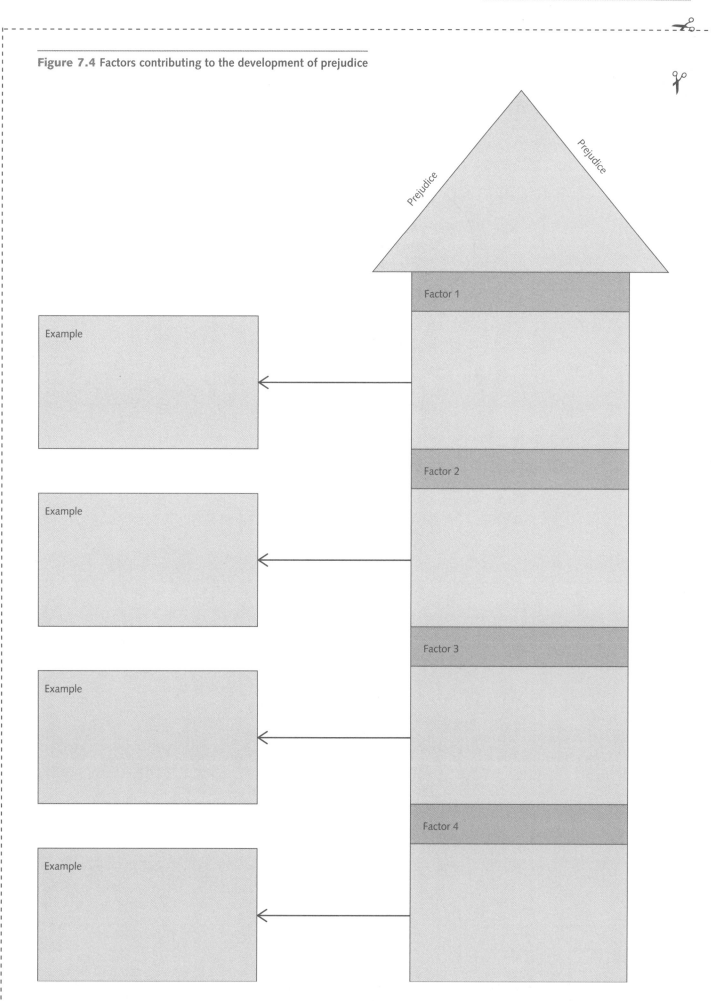

Figure 7.5 Factors contributing to the reduction of prejudice

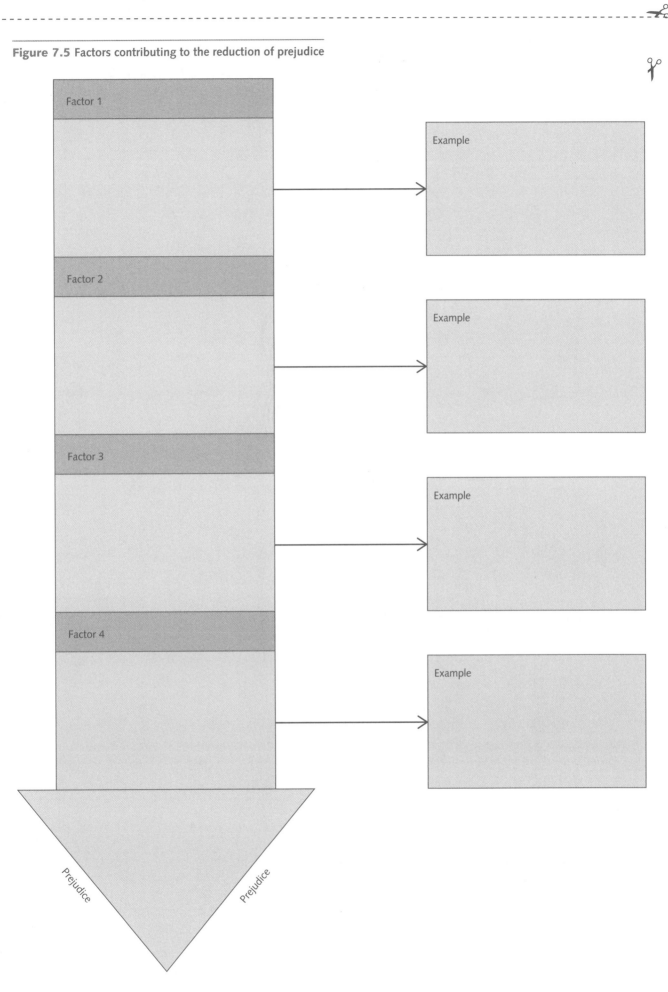

7.4.3 Relationships between attitudes, prejudice and discrimination

Key science skills

Analyse, evaluate and communicate scientific ideas
- discuss relevant psychological information, ideas, concepts, theories and models and the connections between them

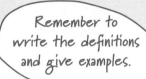

Remember to write the definitions and give examples.

Develop

PART A

1 Study the flowchart in Figure 7.6.

2 Complete the flowchart by writing definitions for attitude, prejudice, discrimination, racism, sexism and ageism in the appropriate boxes.

3 Provide examples of racism, sexism and ageism in the 'Example' rectangles.

PART B

1 Read the four scenarios.

2 Identify where there is evidence of the following terms by underlining that section in each scenario.

- racism

- sexism

- discrimination

- ageism.

3 Describe how these are examples of the term.

Scenario 1

The sports coordinator requests students who are 180 cm (6 feet) tall to try out for a mixed netball competition between colleges. Fiona decides to try out, even though she is 167 cm tall, because she loves to play and has played competitively for years. At after-school try-outs the sports coordinator and sports captains inform her that the team is full and without seeing her perform tell her to 'go on home'. Fiona is very upset and goes home crying.

_____
_____
_____
_____

Scenario 2

Peta arrives 30 minutes early for her job interview. She has made an effort with her appearance and is confident she can do the advertised job of senior sales consultant because she has previous experience at several well-known companies. When the door to the interview office opens, a woman walks out and asks the secretary in the waiting room to 'please send in Peter Walker when he gets here'. Peta stands up and introduces herself. She can tell by the expression on the interviewer's face that something is wrong. During the interview, she feels that the interviewer is uninterested in what she has to say and after just five minutes the interviewer stops the interview by saying 'Look, you just won't do. This job is for the car industry and you being a woman and everything ... sorry, I thought you were a male and your name was Peter.'

_____
_____
_____
_____

Figure 7.6 **The interrelationship between attitudes, prejudice and discrimination**

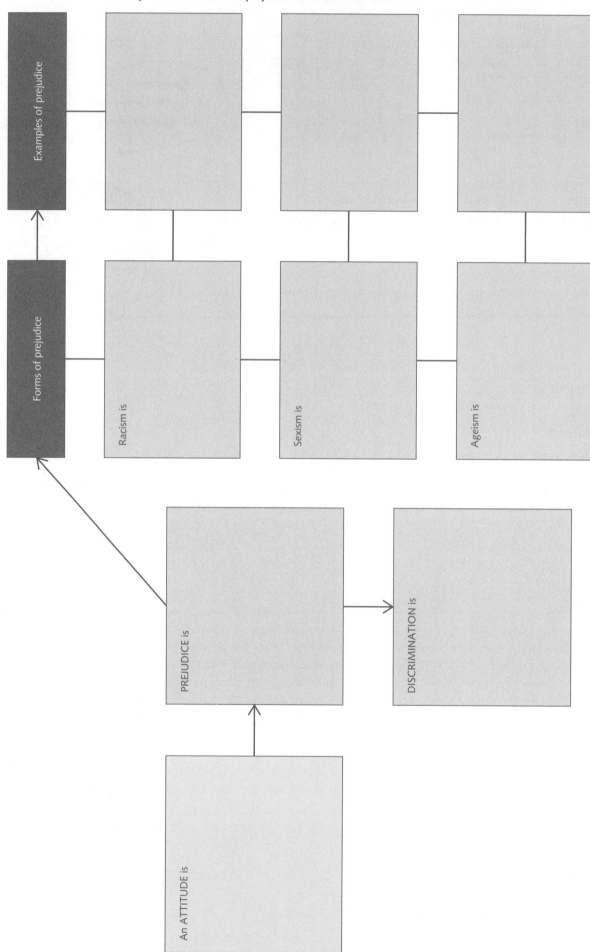

Examples of prejudice

Forms of prejudice

Racism is

Sexism is

Ageism is

PREJUDICE is

DISCRIMINATION is

An ATTITUDE is

Scenario 3

Tai had been a trusted and enthusiastic worker at Hungry Hamburgers for two years. When he turned 18, the manager called him into the office. The manager said, 'Look, now that you're 18, you have to sign a contract and work for us part time because it is too expensive to pay you casual rates'. Tai was a little concerned because he was halfway through his final year of schooling and didn't want to work the minimum 15 hours a week that a part-time contract would include. He explained this to his manager, who said he understood. Later that week when the roster for the next month went up, Tai had only one shift in the entire month instead of the usual three shifts a week.

Scenario 4

Santiago loved watching AFL football and went every week to support his team. He liked to get involved in the game and often would yell out to the players and the referees when he thought they had done something good or made a bad call. One day a man next to him barracking for the opposition team told Santiago to 'sit down and keep quiet. You don't know what you're talking about. Go back to where you came from.' Santiago was a little taken aback by this comment because he had lived all his life in a suburb of Melbourne. When Santiago asked the man what he meant by his comment, the man replied 'You should stick to games more suited to the likes of you, like soccer. AFL is for Australians, not for people like you.'

7.4.4 Prejudice and ways to reduce prejudice

Key science skills

Analyse, evaluate and communicate scientific ideas

- discuss relevant psychological information, ideas, concepts, theories and models and the connections between them
- critically evaluate and interpret a range of scientific and media texts (including journal articles, mass media communications, opinions, policy documents and reports in the public domain), processes, claims and conclusions related to psychology by considering the quality of available evidence

Develop

Read the article and answer the questions that follow.

Muslim women endure a rising wave of violence and intimidation

L. Houlihan, *Herald Sun*
October, 2005

Women wearing head scarves have been spat on, sworn at and assaulted while hate graffiti such as 'Kill Muslims' and 'Muslims out' has appeared in Melbourne's northern suburbs.

In the past two weeks, two Muslim women have been attacked in daylight. In one, a milkshake was hurled at a Muslim woman as she waited at a Sydney Road tram stop with her three children. In the other, a man swerved his car at a Muslim woman and shouted 'F--- off terrorist' while she was crossing the road carrying her baby. There are also reports of Muslim girls being spat at and abused by drivers. The Islamic Council of Victoria is alarmed at the rise of religious abuse.

Coburg resident Omar Merhi was stunned when he saw the car swerve at the woman with her baby. Mr Merhi said Muslim victims often had poor English and felt powerless to take action. Sgt Mick Ehmer, from Brunswick police, encouraged victims to come forward to help stamp out the bigoted behaviour.

'The woman who was hit by a milkshake has been assaulted. We need to be made aware of these (attacks) so we can follow them up and stop them continuing,' he said. Islamic Council of Victoria president, Malcolm Thomas, said anti-Muslim attacks had risen since September 11, 2001. He said the recent spike probably stemmed from the Bali bombings and calls by some politicians for a ban on the Muslim headscarf or hijab.

»

'Comments like that will bring out the rednecks. Most of the time everybody gets along. It's only when the temperature rises due to overseas events that these things happen,' he said. After the most recent Bali bombings, Muslim parents dropping their children off at school were harassed, causing 'a lot of distress', Mr Thomas said.

Melbourne Muslims have been further shaken by recent attacks because they had occurred during their holy month of Ramadan. 'If anyone belittles or insults you, you can only respond by saying "I am fasting",' Mr Merhi said.

Equal Opportunity Commission head, Dr Helen Szoke, said she was appalled by the attacks. 'We recognise that people are fearful about terrorism,' she said. 'But fear cannot be used as an excuse to attack fellow Australians who want nothing more than to go to work, send their kids to school, go to the shops and go about their daily lives without being abused and attacked.'

Source: Houlihan, L., Herald Sun, October 2005.

Questions

1 What type of prejudice is represented in this article?

2 The assault with the milkshake demonstrated which component of an attitude?

3 Identify and explain a factor that may have influenced the formation of such prejudice.

4 Why might Muslim women be the target of these incidents?

5 Suggest a strategy the Equal Opportunity Commission could use in the community to reduce prejudice. In developing your strategy, you should consider the factors that are effective in reducing prejudice, such as superordinate goals, mutual interdependence, cognitive interventions and so on.

7.4.5 Formation and changing of attitudes

Key science skills
Analyse, evaluate and communicate scientific ideas
- discuss relevant psychological information, ideas, concepts, theories and models and the connections between them

Develop

Study the information pairs listed in Table 7.14.

Choose the correct information pair to complete the statements that follow. Write the words in the spaces provided. (Hint: The order of the words in each information pair is the same as the order they appear in the statements.)

Table 7.14 Information pairs

prejudice; discrimination	discrimination; behavioural
racism; ageism	Likert scale; quantitative
mutual interdependence; intergroup contact	affective; emotions
sexism; discrimination	behavioural; actions
prejudice; learned	stereotype; category
sustained contact; equal status contact	media; stereotypes
cognitive interventions; reduce	qualitative; quantitative

1 When we hold a negative emotional attitude towards members of specific social groups based on their membership of that group, we hold a _____. This attitude is _____ and heavily influenced by our interactions with others.

2 _____ refers to an emotional attitude towards others; while _____ refers to the way we behave towards others.

3 According to the tri-component model of attitudes, the _____ component of an attitude refers to the _____ we feel towards the subject of the attitude.

4 Attitude formation is influenced by a wide range of factors, such as the _____. This factor often moulds our attitudes by representing people according to an oversimplified image that causes them to be seen as more similar than they are. These are known as _____.

5 Characteristics of attitudes can be measured by collecting data in written or oral form. This type of data is known as _____ data. If you gathered the data in numerical form and you later used it to statistically describe an attitude characteristic, you would be using _____ data.

6 If you chose not to employ a woman, who meets all your employment criteria, solely on the basis that she might get pregnant in the future, you are guilty of _____, which is a form of _____.

7 One way of reducing prejudice is to have holders of a prejudiced attitude spend time with the target of the prejudiced attitude. However, this method will have a greater chance of reducing prejudice if the interaction continues over a substantial period of time. This is known as _____. If the social interaction involves both parties having the same amount of power or status, then this form of sustained contact is known as _____.

8 If our strategy to avoid prejudice involves making a conscious effort to use our learned skills and behaviour to resist prejudice, we are using the method of _____. This method is one way we can _____ prejudice.

9 A _____ allows you to collect numerical data about attitude characteristics; therefore, it is a form of _____.

10 According to the tri-component model of attitudes, the _____ component of an attitude refers to the _____ we demonstrate towards the subject of our attitude.

11 When we group people together on the basis of perceived similarities, we form a _____ and this places them in a _____ that does not take into consideration individual characteristics of the person.

12 When we treat someone unequally based on our prejudice towards them, we are guilty of _____. According to the tri-component model of attitudes, this reflects the _____ component of an attitude.

13 _____ is a form of prejudice based on race and ethnicity, while _____ is a prejudice based on years of living.

14 One of the cognitive interventions that can be used to combat prejudice involves creating a situation where people have to depend on each other in order to achieve their goal. This strategy is known as _____. Another strategy is to have a long period of interaction between groups. This is known as _____.

Chapter 7 summary

Tell a story about the person in Figure 7.7. Brainstorm your ideas that you have about this person using key psychological concepts as a guide. Your story should include information about their occupation, socioeconomic status, any prejudice or discrimination they may experience, daily activities, personality and intelligence. Compare your answer with someone else's.

Figure 7.7 Image of a person

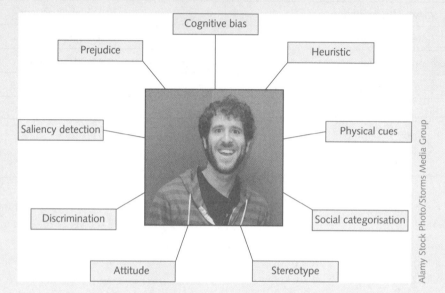

Factors that influence individual and group behaviour

8

Social groups and culture

Key knowledge
- the influence of social groups and culture on individual behaviour

8.1.1 Social power

Key science skills
Analyse, evaluate and communicate scientific ideas
- discuss relevant psychological information, ideas, concepts, theories and models and the connections between them

Develop

This activity will help you understand the factors that affect social power and status.

PART A

1 Define power.

2 Allocate the terms in Table 8.1 to the 'power wheel' in Figure 8.1.

Table 8.1 Our identities accumulate to define our social power

owner	learned English	non-English	poor
large	heterosexual	robust	neurotypical
neuroatypical	average	able bodied	man
gay	mostly stable	slim	some disability
vulnerable	significant disability	significant neurodivergence	lesbian
post-secondary	white	renting	Year 10
citizen	trans	VCE	woman
dark	homeless	middle class	different shades
wealthy	English	work visa	refugee

Figure 8.1 Power wheel – the closer to the middle the more power an individual has

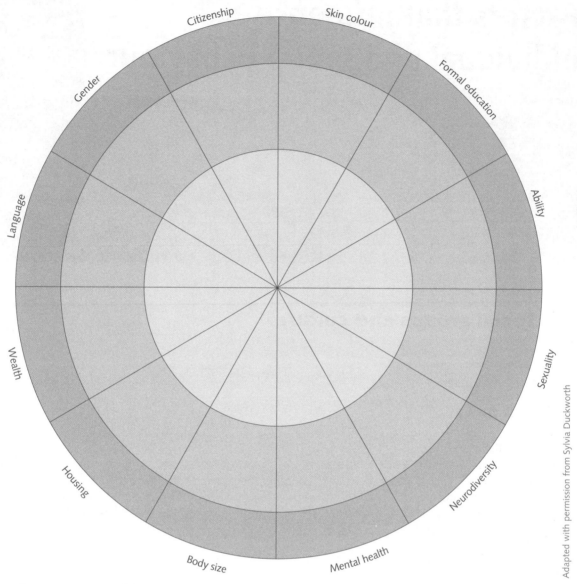

Adapted with permission from Sylvia Duckworth

3   Where do you lie in terms of the power that you have in our society?

_____

_____

PART B

1   Define status.

_____

2   Create a hierarchy from the power wheel (Figure 8.1) of a person with low status in our society.

_____

_____

3   Write a profile of that person.

_____

_____

_____

4   Write a profile of a person with high status within our society.

_____

_____

_____

_____

8.1.2 Social influence

Develop

Read the following scenarios and decide which type of social influence is occurring. Choose from those listed in Table 8.2.

Table 8.2 Social influence

referent power	expert power	coercive power
legitimate power	reward power	

1   You clean your room because your mother tells you to. _____

2   You begin an exercise program because a doctor told you that you needed to._____

3   You get a new haircut because your favourite singer has the same style._____

4   You wear the correct uniform to work because otherwise you will get sent home or fired. _____

5   You try to get the best result in your maths class because the teacher will give the top student a chocolate frog._____

6   Create your own scenarios for each type of social influence listed below

•   Referent

_____

_____

•   Expert

_____

_____

•   Legitimate

_____

_____

•   Coercive

_____

_____

•   Reward

_____

_____

8.1.3 Comparing cultures

Develop

This activity will help you identify the key similarities and differences between individualist and collectivist cultures.

1   Define the following terms:

a. Group

_____

_____

b. Culture

_____

_____

2   Complete the Venn diagram to compare and contrast an individualist culture and a collectivist culture.

Figure 8.2 Comparing features of collectivist and individualist cultures

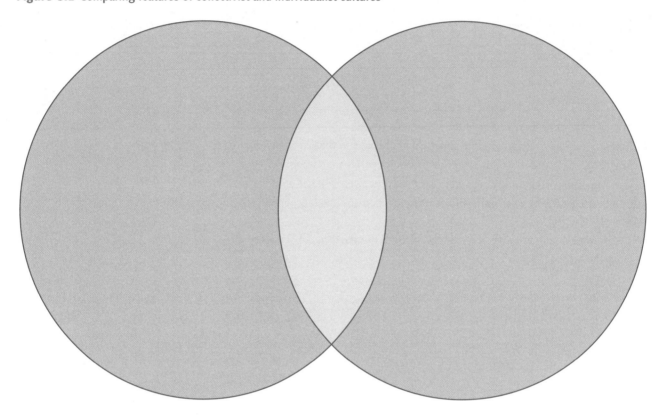

8.1.4 Zimbardo's experiment on the effects of perceived power on behaviour

Key science skills

Develop aims and questions, formulate hypotheses and make predictions
- identify, research and construct aims and questions for investigation
- identify independent, dependent, and controlled variables in controlled experiments
- formulate hypotheses to focus investigation

Comply with safety and ethical guidelines
- demonstrate ethical conduct and apply ethical guidelines when undertaking and reporting investigations

Construct evidence-based arguments and draw conclusions
- identify, describe and explain the limitations of conclusions, including identification of further evidence required

Analyse, evaluate and communicate scientific ideas
- discuss relevant psychological information, ideas, concepts, theories and models and the connections between them

Develop

Complete the flowchart in Figure 8.3 summarising Zimbardo's (1971) Stanford Prison Experiment.

Figure 8.3 A overall summary of the Stanford Prison Experiment

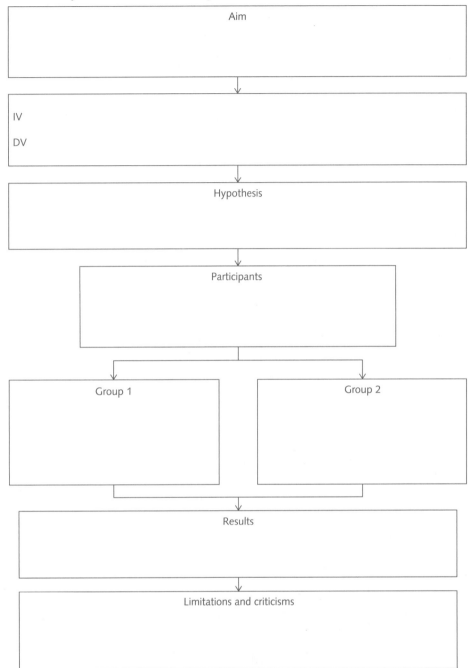

8.2 Obedience and conformity

Key knowledge
- the concepts of obedience and conformity and their relative influence on individual behaviour

8.2.1 Obedience

Key science skills
Analyse, evaluate and communicate scientific ideas
- discuss relevant psychological information, ideas, concepts, theories and models and the connections between them

Develop

MATERIALS

- glue
- scissors

INSTRUCTIONS

1 Cut out the terms and definitions related to obedience in Table 8.3.

2 Match the terms with the definitions, then glue them onto the concept map in Figure 8.4 in the correct places. Make sure you glue the definition and term for obedience in the middle of the diagram.

Figure 8.4 Factors affecting obedience

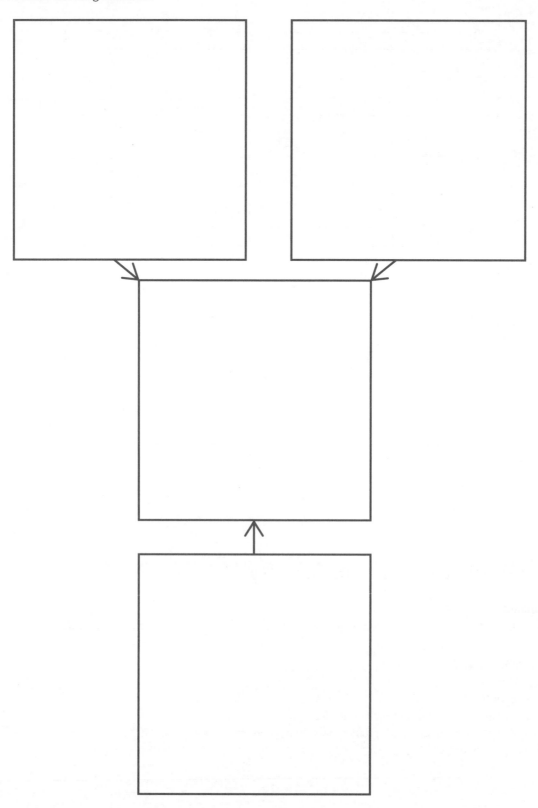

Table 8.3 Terms and definitions related to obedience

Obedience	Social proximity	Group pressure	Legitimacy of authority figures
The social distance between two people; it is more variable than physical distance, because it includes the degree of opportunity to interact.	The higher the status of the person issuing the commands, the more likely a person is to be obedient.	If others in a group are obedient, then a person within that group is more likely to also be obedient.	When people change their behaviour in response to a direct command from someone else.

8.2.2 Milgram's experiment on obedience

Key science skills

Develop aims and questions, formulate hypotheses and make predictions
- identify, research and construct aims and questions for investigation
- identify independent, dependent, and controlled variables in controlled experiments
- formulate hypotheses to focus investigation

Comply with safety and ethical guidelines
- demonstrate ethical conduct and apply ethical guidelines when undertaking and reporting investigations

Construct evidence-based arguments and draw conclusions
- identify, describe and explain the limitations of conclusions, including identification of further evidence required

Analyse, evaluate and communicate scientific ideas
- discuss relevant psychological information, ideas, concepts, theories and models and the connections between them

Develop

In this activity, you will identify the key steps in scientific research undertaken by Milgram (1963).

Read the description of one of Milgram's studies and answer the questions that follow.

Milgram's experiment

Stanley Milgram conducted his first study into obedience in 1962. There were 40 male participants. The study was conducted using two participants at a time. The participants drew slips of paper to see who would take the role of the 'Learner' and who would be the 'Teacher'. An experimenter was present the entire time.

The Learner's task was to memorise a list of word pairs. The Teacher would then test the Learner's memory by reading out the first word in a pair; the Learner would respond with the second word if he could remember it. The Learner was attached to a complex machine and placed in a room, while the Teacher was seated in another room in front of another machine displaying gauges and dials. The Teacher was told to use the machine to deliver a painful electric shock to the Learner each time he gave a wrong answer. The intensity of shocks increased each time a wrong answer was given. The Teacher could hear but not see the Learner; that is, when a shock was delivered, the Teacher could hear the Learner scream. If the Teacher hesitated in giving shocks, the experimenter would advise him that he must continue the experiment.

Although the Teacher believed he was giving electric shocks to the Learner when a mistake was made, no shocks were actually given. Unbeknown to the Teacher, Milgram had told the Learner of the intentions of the study, so the Learner would play along.

The experiment was designed to test how far the Teacher would go in obeying orders when giving the shocks, even if they appeared to be harming the Learner.

1 What was the aim of this experiment?

_____

_____

2 Write a research hypothesis for this experiment.

_____

_____

3 Identify the independent variable(s).

4 Identify the dependent variable(s). _____

5 Who were the participants?

_____

_____

6 Outline the findings of the experiment.

_____

_____

_____

7   Describe how Milgram used deception.

_____

_____

_____

8   Explain the ethical principle of voluntary participation. Was this a factor in this experiment?

_____

_____

_____

9   Outline the debriefing procedure a psychologist must perform at the conclusion of an experiment. Would Milgram have debriefed his participants? What would he have needed to explain to them?

_____

_____

_____

8.2.3 Reproducing Milgram's findings

Key science skills

Analyse and evaluate data and investigation methods
- identify and analyse experimental data qualitatively, applying where appropriate concepts of: accuracy, precision, repeatability, reproducibility and validity; errors; and certainty in data, including effects of sample size on the quality of data obtained

Construct evidence-based arguments and draw conclusions
- discuss the implications of research findings and proposals, including appropriateness and application of data to different cultural groups and cultural biases in data and conclusions

Analyse, evaluate and communicate scientific ideas
- discuss relevant psychological information, ideas, concepts, theories and models and the connections between them
- analyse and explain how models and theories are used to organise and understand observed phenomena and concepts related to psychology, identifying limitations of selected models/theories

Develop

Read the case study and answer the questions that follow.

Researchers have conducted partial and less ethically complicated replications of Milgram's work. As an extension, a group of researchers led by Patrick Haggard, a cognitive neuroscientist, wanted to find out what participants were feeling. They designed a study in which volunteers knowingly inflicted real pain on each other and were completely aware of the experiment's aims.

In these experiments, female volunteers were given $35. In pairs, they sat facing each other across a table, with a keyboard between them. A participant designated the 'agent' could press one of two keys; one did nothing. The other key would transfer five cents to the agent from the other participant, designated the 'victim'. For others, the key would also deliver a painful but bearable electric shock to the victim's arm.

In one experiment, an experimenter stood next to the agent and told her which key to press. In another, the experimenter looked away and gave the agent a free choice about which key to press.

Psychologists have established that people perceive the interval between an action and its outcome as shorter when they carry out an intentional action of their own free will, such as moving their arm, than when the action is passive, such as having their arm moved by someone else.

To examine the participants' 'sense of agency' – the unconscious feeling that they were in control of their own actions – Haggard and his colleagues designed the experiment so that pressing either key caused a tone to sound after a few hundred milliseconds, and both volunteers were asked to judge the length of this interval.

When they were ordered to press a key, the participants seemed to judge their action as more passive than when they had free choice – they perceived the time to the tone as longer.

In a separate experiment, female volunteers followed similar protocols while electrodes on their heads recorded their neural activity through EEG (electroencephalography). When ordered to press a key, their EEG recordings were quieter — suggesting,

»

says Haggard, that their brains were not processing the outcome of their action. Some participants later reported feeling reduced responsibility for their action.

Unexpectedly, giving the order to press the key was enough to cause the effects, even when the keystroke led to no physical or financial harm. It appears as if someone's sense of responsibility is reduced whenever someone orders another person to do something, whatever it is they are telling them to do.

source: https://www.nature.com/articles/nature.2016.19408

1 What does this investigation tell you about the reasons we obey others?

_____

_____

2 Haggard only used female participants. What was his reason for doing this?

_____

_____

3 How did Haggard measure 'sense of agency'?

_____

_____

4 How could you apply these findings in a school setting?

_____

_____

_____

5 Define reproducibility. How does this investigation fit with this definition?

_____

_____

_____

_____

8.2.4 Conformity

Key science skills

Analyse, evaluate and communicate scientific ideas
* discuss relevant psychological information, ideas, concepts, theories and models and the connections between them

Develop

MATERIALS

* glue
* scissors

INSTRUCTIONS

1 Cut out the terms and definitions related to conformity in Table 8.4.
2 Match the terms with the definitions and glue them onto the concept map in Figure 8.5 in the correct places. Make sure you glue the definition and term for 'conformity' in the middle of the diagram.

Figure 8.5 Factors affecting conformity

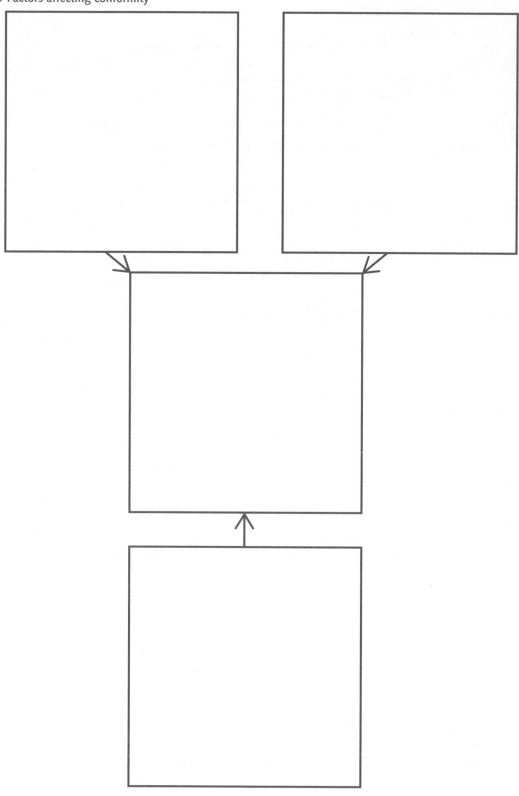

Table 8.4 Terms and definitions related to conformity

Conformity	Normative influence	Culture	Informational influence
Group size	Unanimity	Deindividuation	Social loafing
An agreement among all members of a group.	The impact of the standards established by the group on an individual's behaviour.	The larger the group, the more likely a person is to conform.	Where the social environment or situational cues are used by individuals to help them monitor or adapt their behaviour to fit with the behaviour of those around them.
The tendency for an individual to reduce their effort in a group.	When individuals change their behaviour in response to real or implied pressure by others.	The loss of social identity and inhibition, causing a person to lose responsibility for their own actions and to ignore possible consequences.	Whether a person belongs to an individualist or collectivist culture.

8.2.5 Social factors that affect group decision making

Key science skills

Analyse, evaluate and communicate scientific ideas
- discuss relevant psychological information, ideas, concepts, theories and models and the connections between them

Develop

The following activity will help you consolidate your understanding of the factors that influence group decision making.

1 Define Groupthink.

_____

2 Match the terms in Table 8.5 to their effects in Table 8.6.

Table 8.5 Terms used to describe Groupthink

invulnerability	stereotypes	unanimity
rationale	pressure	mindguards
morality	self-censorship	

Table 8.6 Effects of Groupthink on decision making

Symptom of Groupthink	Effect on the group
	These people filter out negative information to protect the decision of the group
	A justification that the group decision is the right one regardless of consequences
	Members believe that the decisions of the group are shared by all
	Applying persuasion or coercion to any member that challenges a group decision
	Members avoid challenging the group decision
	The silence of members during a discussion is interpreted as support
	Discounting the opinions of other groups by referring to them in a negative light
	Discounting evidence that opposes a group decision
	An arrogance that results in risk-taking decisions

8.2.6 True or false?

Key science skills

Analyse, evaluate and communicate scientific ideas
- discuss relevant psychological information, ideas, concepts, theories and models and the connections between them

Develop

Indicate whether each statement is true or false by placing a tick in the correct column in Table 8.7.

Table 8.7 True or false statements about social influence

Statement	True	False
Our behaviour is the same regardless of whether we are alone or in a group.		
A group is two or more people interacting with a shared purpose.		
You are more likely to follow a well-dressed person across the road with a red walk sign than a person who is poorly dressed.		
The Stanford Prison Experiment (Zimbardo, 1971) demonstrated that prison guards are naturally aggressive.		
People have only one type of power.		
Doing what your parents tell you to do is an example of obedience.		
In Milgram's experiment (1963), the 'Learner' was really a confederate.		
The gradual nature of Milgram's experiment (1963) was thought to make no difference to the shock level delivered by participants.		
At work, you are more likely to obey a manager than to obey a colleague.		
In Asch's study (1951), it was difficult to tell which lines were of the same length.		
The dominant culture in Australia is best described as individualistic.		
The size of the group does not affect the likelihood of conformity.		
In very large groups or crowds, most people do not accept responsibility for their own behaviour.		
The UK is an example of a vertical-collectivist culture.		
Taking your shoes off at a friend's place the first time you visit is an example of informational influence.		
The Stanford Prison Experiment (Zimbardo, 1971) is an example of researcher bias.		
Our willingness to conform protects us from ridicule.		

8.3 Media and mental wellbeing

Key knowledge
- positive and negative influences of different media sources on individual and group behaviour such as changing nature of social connections, social comparison, addictive behaviours and information access

8.3.1 Social comparison scale

Develop

Key science skills

Develop aims and questions, formulate hypotheses and make predictions
- identify, research and construct aims and questions for investigation
- formulate hypotheses to focus investigation

Comply with safety and ethical guidelines
- demonstrate ethical conduct and apply ethical guidelines when undertaking and reporting investigations

Generate, collate and record data
- systematically generate and record primary data, and collate secondary data, appropriate to the investigation

Analyse and evaluate data and investigation methods
- process quantitative data using appropriate mathematical relationships and units, including calculations of percentages, percentage change and measures of central tendencies (mean, median, mode), and demonstrate an understanding of standard deviation as a measure of variability

Construct evidence-based arguments and draw conclusions
- evaluate data to determine the degree to which the evidence supports or refutes the initial prediction or hypothesis

PART A

Complete the self-report measure of social comparison in Table 8.8.

Table 8.8 Social comparison Likert scale

Most people compare themselves from time to time with others. For example, they may compare the way they feel, their opinions, their abilities, and/or their situation with those of other people. There is nothing particularly 'good' or 'bad' about this type of comparison, and some people do it more than others. To find out how often you compare yourself with other people, place a number next to the statement to indicate how much you agree with each statement, by using the following scale.

1	2	3	4	5
I disagree strongly				**I agree strongly**

1	I often compare how my loved ones (boyfriend or girlfriend, family members, etc.) are doing with how others are doing. _____
2	I always pay a lot of attention to how I do things compared with how others do things.
3	If I want to find out how well I have done something, I compare what I have done with how others have done.
4	I often compare how I am doing socially (e.g. social skills, popularity) with other people.
5	I am not the type of person who compares often with others. (reversed)
6	I often compare myself with others with respect to what I have accomplished in life.
7	I often like to talk with others about mutual opinions and experiences.
8	I often try to find out what others think who face similar problems as I face.
9	I always like to know what others in a similar situation would do.
10	If I want to learn more about something, I try to find out what others think about it.
11	I never consider my situation in life relative to that of other people. (reversed)

Source: The Questionnaire of the Iowa-Netherlands Comparison Orientation Measure proposed by Gibbons and Buunk (1999).

Adapted from: Schneider, S., & Schupp, J. (2011). The Social Comparison Scale: Testing the Validity, Reliability, and Applicability of the Iowa-Netherlands Comparison Orientation Measure (INCOM) on the German Population. DIW Data Documentation. 55.

SCORING

Add up your scores. For items 11 and 5, however, reverse your scoring. If you chose 1 for item 11, for example, instead score 5. The higher the score, the more prone you are to compare yourself with others.

PART B

Replicate the survey from Part A with different year levels or genders in your school to determine any patterns or trends, then answer the questions that follow.

Table 8.9 Average score of social comparison across year levels

Year level	Average score
7	
8	
9	
10	
11	
12	

Table 8.10 Average score of social comparison across gender

Gender	Average score
Male	
Female	
Other	

1 What is your research question?

2 What is your aim?

3 What do you hypothesise?

4 What would be your ethical considerations?

5 Describe your results.

6 Did your results support or refute your hypothesis?

7 Using your understanding of social influence, explain any interesting trends or results.

8.3.2 Identifying and applying social comparison theory

Key science skills
Construct evidence-based arguments and draw conclusions
- use reasoning to construct scientific arguments, and to draw and justify conclusions consistent with evidence base and relevant to the question under investigation

Analyse, evaluate and communicate scientific ideas
- discuss relevant psychological information, ideas, concepts, theories and models and the connections between them
- analyse and explain how models and theories are used to organise and understand observed phenomena and concepts related to psychology, identifying limitations of selected models/theories

Develop

PART A

For each of the comments in Table 8.11 assign whether the person is making an 'upward social comparison', 'lateral social comparison' or a 'downward social comparison' by writing the letters U, L or D in the right-hand column.

Table 8.11 Examples of social comparison

	Social comparison (U, L or D
My neighbour inspires me. If he can run a half-marathon, then so can I.	
My friend has saved enough to buy her first car. If she can, then so can I.	
My friend is smarter than I am. She just gets it.	
That group of friends over there never fight. My friends always fight; I want to be in a friendship group like that.	
I want to work hard so that I can earn more money than my boss.	
I want to train really hard so that I can make the state team.	
I feel happy knowing that I beat my neighbour in the half-marathon.	
When I see people fight, I'm reminded to be grateful for my relationships.	
I want to travel the world like my favourite influencers on Instagram.	
My disorganised classmate reminds me to plan my work better so that I'm not in the same position that they're in.	
My body shape will never fit the clothes that all the models are wearing.	
My classmate struggles all the time with the same topics, whereas it just clicks for me.	
I want to play the drums like Dave Grohl of the Foo Fighters.	
I just want to check the test scores of my classmates.	

PART B

Determine the effects that social comparison can have on people's thoughts and behaviour.

1 Choose the words in Table 8.12 and write them in the appropriate box in Table 8.13. You can use the terms more than once.

2 Provide an example from your own or others' experience that demonstrates this effect.

Table 8.12 Common effects that result from social comparison

hope	envy
unrealistic	inspiration
scorn	motivation
dissatisfaction	gratitude
reassurance	self-protection

Table 8.13 Positive and negative effects of social comparison

Effect	Upward comparison	Lateral comparison	Downward comparison
Positive			
Example			
Negative			
Example			

PART C

Analyse each of the ads in the table below to determine how advertisers use social comparison theory to sell their product.

Table 8.14 Social comparison theory and advertising

Example	Type of comparison and your justification
Tommy Hilfiger giving his clothes to hip-hop stars to wear on stage 	
Sportswear companies giving clothes to famous athletes 	
Superannuation companies wanting to sell more product 	

8.4 Independence and anti-conformity

Key knowledge
• the development of independence and anti-conformity to empower individual decision making when in groups

8.4.1 Techniques to empower individual decision making

Key science skills

Develop aims and questions, formulate hypotheses and make predictions
• identify, research and construct aims and questions for investigation
• formulate hypotheses to focus investigation

Comply with safety and ethical guidelines
• demonstrate ethical conduct and apply ethical guidelines when undertaking and reporting investigations

Analyse and evaluate data and investigation methods
• process quantitative data using appropriate mathematical relationships and units, including calculations of percentages, percentage change and measures of central tendencies (mean, median, mode), and demonstrate an understanding of standard deviation as a measure of variability

Construct evidence-based arguments and draw conclusions
• evaluate data to determine the degree to which the evidence supports or refutes the initial prediction or hypothesis
• discuss the implications of research findings and proposals, including appropriateness and application of data to different cultural groups and cultural biases in data and conclusions

Develop

PART A

Using the Internet, research each of the techniques listed in Table 8.15 and complete the table adding a description and explanation for each.

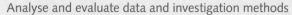

Table 8.15 Techniques to reduce Groupthink

Technique	Description	How it works to reduce the effects of Groupthink within groups
Brainstorming		
De Bono's Six Hat Thinking		
The Delphi Technique		
Modified Borda Count		

PART B

Conduct a study where you compare the outcome of a decision using traditional group discussion with a technique that overcomes the effects of conformity.

Topics can be as benign as which brand of t-shirt is the best or you may choose a more sensitive topic such as the removal of historical monuments because of the negative effect they have on members of our communities.

1 What is your research question?

_____

_____

2 What is your aim?

_____

_____

3 What do you hypothesise?

_____

_____

4 What would be your ethical considerations?

_____

_____

_____

5 Describe your results.

_____

_____

_____

_____

6 Did your results support or refute your hypothesis?

_____

_____

Chapter 8 summary

Use the clues provided to complete the crossword in Figure 8.6.

Figure 8.6 Crossword of terms for social influences on the individual

Across

5 This culture places the needs of the individual above the interests or group

7 Upward social comparison, downward social comparison and lateral social comparison

8 A deliberate refusal to comply with social norms as demonstrated by ideas, beliefs or judgements that challenge said social norms

10 The experience of feeling close and connected to others in a group

11 The established norms (rules) a group follows

12 Formal authority enables an individual to extent influence as a result of either high, legally recognised office

13 Involves resisting the influence of the group

14 To actively argue against the points of group members, even if they share the same beliefs of the group members

15 The loss of social identity and Inhibition when in a group, which results in a person engaging in behaviours with little regard for potential consequences

17 This culture values group needs or interests over the interests of the individual

20 People change their behaviour in response to direct commands from others.

Down

1 The tendency for group members discussing an issue or dilemma to adopt a more extreme position than their individual position prior to the discussion

2 The amount of influence that an individual can exert over another person

3 When individuals change their behaviour as the result or real or implied pressure from others

4 When two or more people interact and influence each other and share a common objective

5 When we look to others' behaviour to guide our own behaviour

6 A person's position in the hierarchy or a group

9 Phenomenon that occurs when a members of a group prioritise the cohesion of the group over objective decision making

16 The expectations and behaviours associated with a social group

18 An individual's sense of self defined by a set of physical, psychological and interpersonal characteristics that is not wholly shared with any other person

19 Complete agreement among every member of group

9 Perception

9.1 Making sense of the world

9.1.1 Sensation

Develop

PART A

1   Using your understanding of sensation, complete Tables 9.1 and 9.2.

Table 9.1 Organs involved in the reception of stimulus energy

Stimulus energy	Sense	Sensory organ
Electromagnetic radiation (light)	Sight (vision)	Eye
	Hearing (audition)	
		Nose
Chemical energy (molecules)		
	Skin (touch, pressure, pain, cold and warmth)	

2 For each sensory organ in Table 9.2, determine the type of sensory receptor cell/s that are stimulated by the stimulus energy.

Table 9.2 Receptor cells within sensory organs

Sensory organ	Sensory receptor cells
Eye	Retina (cones and rods)
Nose	

PART B

Label the key structures of the eye that take light energy and convert it into visual sensation, and describe their function.

Figure 9.1 Structures involved in visual sensation

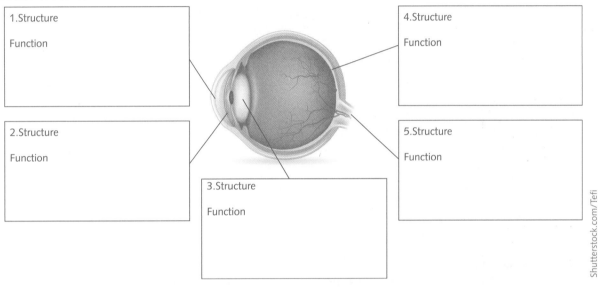

Shutterstock.com/Tefi

1. Structure

Function

2. Structure

Function

3. Structure

Function

4. Structure

Function

5. Structure

Function

PART C

Label the structures of the tongue that turn chemical energy (molecules) into taste sensation and describe their function.

Figure 9.2 Structures involved in taste sensation

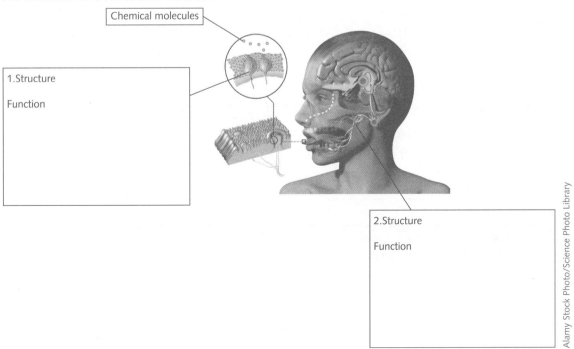

Chemical molecules

1. Structure

Function

2. Structure

Function

Alamy Stock Photo/Science Photo Library

9.1.2 Sensation and perception

Key science skills

Analyse, evaluate and communicate scientific ideas

- discuss relevant psychological information, ideas, concepts, theories and models and the connections between them
- analyse and explain how models and theories are used to organise and understand observed phenomena and concepts related to psychology, identifying limitations of selected models/theories

Develop

1 Study the flowchart of sensation and perception in Figure 9.3.

2 Name the stages in sensation and perception and answer the questions in the spaces provided. Note: Stage 1 is complete and can be used as a model.

Figure 9.3 A sensation and perception flowchart

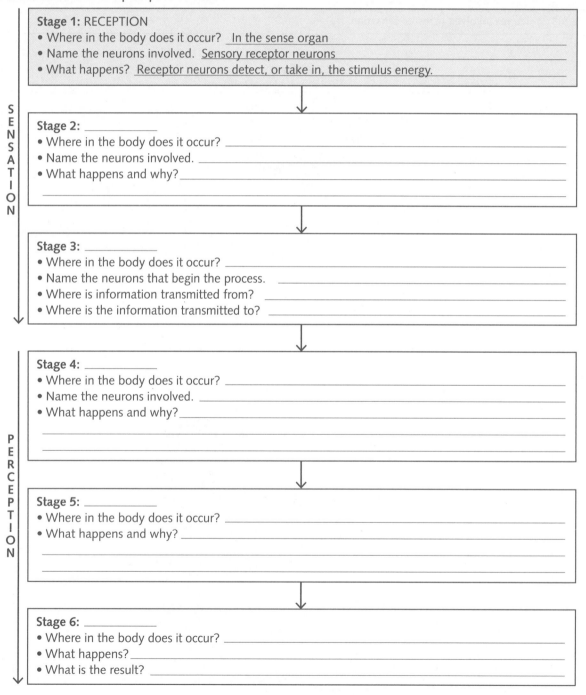

SENSATION

Stage 1: RECEPTION
- Where in the body does it occur? In the sense organ
- Name the neurons involved. Sensory receptor neurons
- What happens? Receptor neurons detect, or take in, the stimulus energy.

Stage 2: _____
- Where in the body does it occur? _____
- Name the neurons involved. _____
- What happens and why? _____

Stage 3: _____
- Where in the body does it occur? _____
- Name the neurons that begin the process. _____
- Where is information transmitted from? _____
- Where is the information transmitted to? _____

PERCEPTION

Stage 4: _____
- Where in the body does it occur? _____
- Name the neurons involved. _____
- What happens and why? _____

Stage 5: _____
- Where in the body does it occur? _____
- What happens and why? _____

Stage 6: _____
- Where in the body does it occur? _____
- What happens? _____
- What is the result? _____

9.1.3 Bottom-up and top-down processing

Key science skills

Analyse, evaluate and communicate scientific ideas

- discuss relevant psychological information, ideas, concepts, theories and models and the connections between them
- analyse and explain how models and theories are used to organise and understand observed phenomena and concepts related to psychology, identifying limitations of selected models/theories

Develop

PART A

Select terms from Table 9.3 to complete the passage about the differences between bottom-up and top-down processing.

Table 9.3 Terms to help describe bottom-up and top-down processing

incoming	understanding	unfamiliar
tiny details	categorisation	general
schemas	new	interpretation
attend	blanks	lost
guess	highly	memories
whole	prior knowledge	know

1 Bottom-up processing is information processing in which _____ sensory stimuli initiate and determine the higher-level processes involved in their _____ and _____.

2 Perception starts with the processing of _____ _____ of sensory stimuli (such as lines, edges, shape and colour).

3 More and more analysis occurs as we build upwards towards a complete _____, or perception, of what the original stimulus represents.

4 We use bottom-up processing when we are presented with _____ or _____ complex stimuli.

5 Top-down processing starts with perceiving the _____ object. You begin with the most _____ features and move towards the more specific or smaller details.

6 We perceive the world around us by drawing from what we already _____ in order to interpret new information. We are not able to _____ to each sensation and most of it is _____ by the time it reaches the brain. To fill in the _____, we formulate an 'educated _____' about how to interpret a particular pattern of sensory stimulation) and anticipate what's next.

7 To do this we rely on _____, which are concepts that include _____ and _____. We use these to recognise patterns in _____ stimuli so we can rapidly form a meaningful perception.

PART B

View each of the images in Table 9.4, then try and figure out what you perceived first, to decide whether you engaged 'top-down' or 'bottom-up' processing. You may want to test this with a larger group of participants.

Table 9.4 How are images perceived?

Image	Top-down or bottom-up? Justify.

Image sources: K. M. D. "A Puzzle-Picture with a New Principle of Concealment." The American Journal of Psychology, vol. 64, no. 3, 1951, pp. 431–33. JSTOR, https://doi.org/10.2307/1419008. Accessed 12 Oct. 2022; Shutterstock.com/file404; Teufel, C., Dakin, S.C. & Fletcher, P.C. Prior object-knowledge sharpens properties of early visual feature-detectors. Sci Rep 8, 10853 (2018). https://doi.org/10.1038/s41598-018-28845-5. Licensed under Creative Commons 4.0., http://creativecommons.org/licenses/by/4.0/

9.2 The role of attention in perception

9.2.1 Understanding attention

PART A

Decide whether the attention would be mainly 'internal' or 'external' in each of these situations by writing 'I' or 'E' in the column on the right in Table 9.5.

Table 9.5 Attention: Internal or External?

Situation	I or E
Experiencing a toothache	
Hearing the sound of a dentist's drill	
Attending a dance club but feeling very anxious	
Reading a book on a sunny day at the beach	
Hearing your name being called in a loud café	
Being aware of your heartbeat	
Experiencing the sensation that someone is watching you	

PART B

Decide whether each situation would require or demonstrate 'sustained', 'divided' or 'selective' attention by writing the type in the right-hand column in Table 9.6.

Table 9.6 Sustained, divided or selective attention?

Situation	Type of attention
Driving to a friend's party and talking to one of your passengers	
Preparing your lunch and organising to be picked up after basketball practice	
Studying for an exam and not realising how much time has passed	
Practising goal kicking	
Studying for an exam and checking your phone	
Having a conversation with a friend in a crowded place	
Being a lifeguard at the local pool	
Designing an ad in vis comm with a specific focal point	

»

Situation	Type of attention
Walking along a narrow brick fence to get from point A to point B without falling off	
Driving up to Mount Buller to go skiing	
Daydreaming in class and hearing your name being called to answer a question	
Performing a dance piece at the end-of-year concert	
Experiencing a toothache while completing a physics test	
Reading a book on a sunny beach	

PART C

Determine whether these situations require or demonstrate an 'automatic' or 'controlled' process and fill in your answer in the right-hand column in Table 9.7.

Table 9.7 Automatic or controlled process?

Situation	Type of processing
Learning to shift gears during your first driving lesson	
Recognising a friend's face in a crowd	
Stacking supermarket shelves	
Learning definitions from your psychology textbook	
Finding your way to the toilet in the middle of the night	
Shifting gears after 5 years of driving	
Reading a book on a sunny beach	
Responding to a feint during a soccer match	

PART D

This activity is an example of a task that is used during rehabilitation to exercise elements of cognition, in this case memory and attention.

1 Study the pattern sets in Figure 9.4 by comparing the left and right columns.
2 Circle where the pattern on the right is different from the pattern on the left. If both sets match, move to the next set of patterns.

Figure 9.4 Same or different – a rehabilitation game testing attention

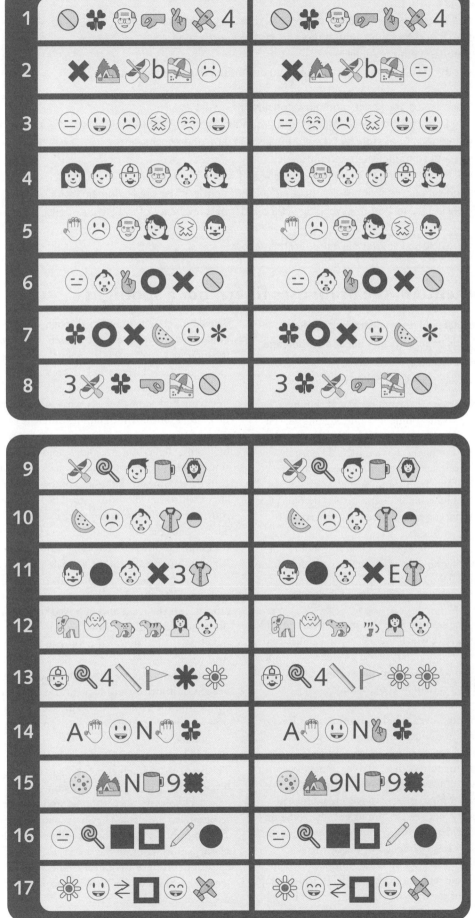

9.2.2 Media analysis

Key science skills

Develop aims and questions, formulate hypotheses and make predictions
- identify, research and construct aims and questions for investigation

Comply with safety and ethical guidelines
- demonstrate ethical conduct and apply ethical guidelines when undertaking and reporting investigations

Construct evidence-based arguments and draw conclusions
- discuss the implications of research findings and proposals, including appropriateness and application of data to different cultural groups and cultural biases in data and conclusions

Analyse, evaluate and communicate scientific ideas
- critically evaluate and interpret a range of scientific and media texts (including journal articles, mass media communications, opinions, policy documents and reports in the public domain), processes, claims and conclusions related to psychology by considering the quality of available evidence

Develop

Read the following article and answer the questions that follow.

To pay attention, the brain uses filters, not a spotlight

A brain circuit that suppresses distracting sensory information holds important clues about attention and other cognitive processes

Jordana Cepelewicz, 24 September 2019

Attentional processes are the brain's way of shining a searchlight on relevant stimuli and filtering out the rest. Neuroscientists want to determine the circuits that aim and power that searchlight. For decades, their studies have revolved around the cortex, the folded structure on the outside of the brain commonly associated with intelligence and higher-order cognition. It's become clear that activity in the cortex boosts sensory processing to enhance features of interest.

But now, some researchers are trying a different approach, studying how the brain suppresses information rather than how it augments it.

Hunting for circuits

Francis Crick, the scientist involved in researching DNA, developed a theory in which the sensory thalamus acted not just as a relay station, but also as a gatekeeper – not just a bridge, but a sieve – staunching some of the flow of data to establish a certain level of focus.

But decades passed, and attempts to identify an actual mechanism proved less than fruitful – not least because of how enormously difficult it is to establish methods for studying attention in lab animals.

That didn't stop Michael Halassa, a neuroscientist at the McGovern Institute for Brain Research at the Massachusetts Institute of Technology. He wanted to determine exactly how sensory inputs got filtered before information reached the cortex, to pin down the precise circuit that Crick's work implied would be there.

He was drawn to a thin layer of inhibitory neurons called the thalamic reticular nucleus (TRN), which wraps around the rest of the thalamus like a shell. The TRN

seemed to let sensory inputs through when an animal was awake and attentive to something in its environment, but it suppressed them when the animal was asleep.

In 2015, Halassa and his colleagues discovered another, finer level of gating – this time involving how animals select what to focus on when their attention is divided among different senses. In the study, the researchers used mice trained to run as directed by flashing lights and sweeping audio tones. They then simultaneously presented the animals with conflicting commands from the lights and tones, but also cued them about which signal to disregard. The mice's responses showed how effectively they were focusing their attention. Throughout the task, the researchers used well-established techniques to shut off activity in various brain regions to see what interfered with the animals' performance.

As expected, the prefrontal cortex, which issues high-level commands to other parts of the brain, was crucial. But the team also observed that if a trial required the mice to attend to vision, turning on neurons in the visual TRN interfered with their performance. And when those neurons were silenced, the mice had more difficulty paying attention to sound. If the mouse needed to prioritize auditory information, the prefrontal cortex told the visual TRN to increase its activity to suppress the visual thalamus – stripping away irrelevant visual data.

The attentional searchlight metaphor was backward: The brain wasn't brightening the light on stimuli of interest; it was lowering the lights on everything else.

Despite the success of the study, the researchers recognized a problem. They had confirmed Crick's hunch: The prefrontal cortex controls a filter on incoming sensory information in the thalamus. But the prefrontal cortex doesn't have any direct connections to the sensory portions of the TRN. Some part of the circuit was missing.

»

Until now. Halassa and his colleagues have finally put the rest of the pieces in place, and the results reveal much about how we should be approaching the study of attention.

With tasks similar to those they'd used in 2015, the team probed the functional effects of various brain regions on one another, as well as the neuronal connections between them. The full circuit, they found, goes from the prefrontal cortex to a much deeper structure called the basal ganglia (often associated with motor control and a host of other functions), then to the TRN and the thalamus, before finally going back up to higher cortical regions. So, for instance, as visual information passes from the eye to the visual thalamus, it can get intercepted almost immediately if it's not relevant to the given task. The basal ganglia can step in and activate the visual TRN to screen out the extraneous stimuli, in keeping with the prefrontal cortex's directive.

Furthermore, the researchers found that the mechanism doesn't just filter out one sense to raise awareness of another: It filters information within a single sense too. When the mice were cued to pay attention to certain sounds, the TRN helped to suppress irrelevant background noise within the auditory signal.

Halassa's findings indicate that the brain casts extraneous perceptions aside earlier than expected. 'What's interesting,' said Ian Fiebelkorn, a cognitive neuroscientist at Princeton University, is that 'filtering is starting at that very first step, before the information even reaches the visual cortex.'

Halassa's discovery of the basal ganglia's role in attention is particularly fascinating. That's partly because it is such an ancient area of the brain, one that hasn't typically been viewed as a part of selective attention. 'Fish have this,' Krauzlis said. 'Going back to the earliest vertebrates, like the lamprey, which doesn't have a jaw' – or a neocortex, for that matter – 'they have basically a simple form of basal ganglia and some of these same circuits.' The fishes' neural circuitry may offer hints about how attention evolved.

Halassa is particularly intrigued by what the connection between attention and the basal ganglia might reveal about conditions like attention deficit hyperactivity disorder and autism, which often manifest as hypersensitivity to certain kinds of inputs.

This passage is adapted from Jordana Cepelewicz's "To Pay Attention, the Brain Uses Filters, Not a Spotlight," originally published in Quanta Magazine, an editorially independent online publication supported by the Simons Foundation.

1 What was the aim of Halassa's 2015 investigation?

_____

2 What ethical considerations would Halassa have to implement for this investigation?

_____

_____

3 Label the location of the thalamus, the basal ganglia, the frontal lobe and the visual cortex on the diagram in Figure 9.5.

Figure 9.5 Key structures involved in attention

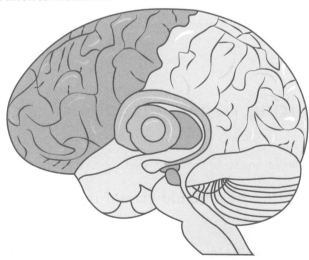

4 Annotate Figure 9.5 with one or more crosses (X) where the TRN (thalamic reticular nucleus) network is located.

5 In the box provided, draw a neural circuit that allows a student to focus on writing notes from the board, in a room where other students are walking around completing other tasks. Use the following terms: light stimulus, retina, optic nerve, prefrontal cortex, basal ganglia, TRN, thalamus and visual cortex.

6 What is ADHD?

7 What are the biological factors that influence the expression of ADHD?

8 How could the research outlined above be applied to our understanding of ADHD and perhaps how to treat the disorder?

9 What features are contained within this article that give it authority?

Factors influencing visual perception

Key knowledge
- the influence of biological, psychological and social factors on visual perception and gustatory perception

9.3.1 Sensation and perception of a visual stimulus

Key science skills
Analyse, evaluate and communicate scientific ideas
- discuss relevant psychological information, ideas, concepts, theories and models and the connections between them
- analyse and explain how models and theories are used to organise and understand observed phenomena and concepts related to psychology, identifying limitations of selected models/theories

Develop

PART A

This activity will reinforce your understanding of the biological components involved in visual sensation and perception.

1 Study Figure 9.6.

2 Using bullet points, explain what each stage of visual sensation and perception contributes to your experience of knowing that the visual stimulus you are looking at is a koala.

Figure 9.6 What contributes to your knowing this is a koala?

iStock.com/GlobalP

_____

_____

_____

_____

_____

_____

_____

3 Study the information in the list that follows.

- Transduces light into impulses of electrochemical energy
- Assigns meaning to the stimulus
- Feature-detector cells select the most important information and ignore the rest
- Receives the light stimulus
- Is a gap in the retina where stimulus leaves
- Transmits stimulus from the retina to the brain
- Is electromagnetic energy
- Uses visual perception principles to organise the stimulus into a new form that can be interpreted.

4 Write each point in the correct spot in Figure 9.7.

Figure 9.7 The pathway for vision

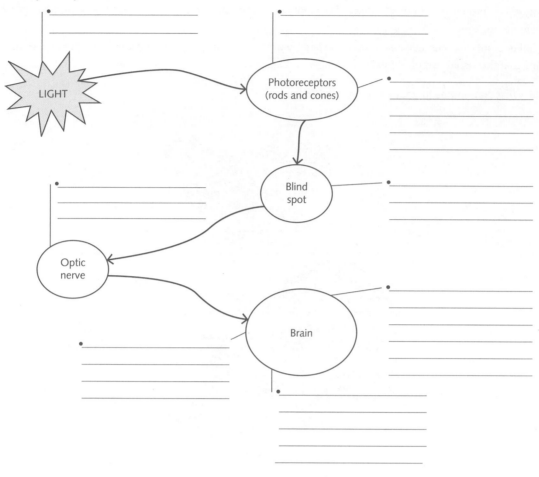

PART B

Complete Table 9.8 to identify the key characteristics of photoreceptors.

Table 9.8 The key characteristics of photoreceptors

	Cones	Rods
Location in the retina		
How many there are		
Light conditions for optimal functioning		
Type of vision they are responsible for		

PART C

Complete the paragraphs about the process of visual perception by writing the terms from Table 9.9 in the correct spaces.

Table 9.9 Terms associated with visual perception

visual perception	reception	optic nerve
cones	electrochemical energy	visual perception principles
organisation	transmission	visual sensation
electromagnetic energy	selection	rods
interpretation	light	transduction
light waves	visual cortex	feature detectors

The process by which visual information is registered by our visual sense and transmitted to the brain is known as _____. If you walked into your lounge room and found a giraffe standing there, you would see the giraffe partly because receptors in your eyes absorb _____, which are a form of _____. When the retina's photoreceptors absorb enough electromagnetic energy (more commonly called _____) to activate them, we refer to this as _____. If it was very dark, the photoreceptors registering the image of the giraffe would be the _____, so you would not see the contrasting colours of the giraffe's skin or any fine detail of its build. However, if all the lights were on, your _____ would register this image in all its colour and detail.

Your photoreceptors would convert the giraffe's image into nerve impulses of _____. This process is known as _____. The impulses would travel via your _____ to the _____ of your brain, in a process known as _____. Not all of the impulses that reach your visual cortex would be processed because specialised neurons, known as _____, only respond to certain features of the information and ignore the rest. This part of the process is known as _____.

Your brain now has to reassemble the filtered stimuli into a form or pattern that it can recognise. This occurs during _____, when the brain automatically applies a set of rules, known as _____, to help it quickly organise the stimuli in a consistent and meaningful way. Once this has occurred, your brain is able to give meaning to the original light stimulus. This means that your brain has completed the final stage of processing the visual information and this is called _____. Now your brain knows that the light image your eyes received is an image of a giraffe, so you have experienced a _____.

9.3.2 Visual perception

Key science skills
Analyse, evaluate and communicate scientific ideas
- discuss relevant psychological information, ideas, concepts, theories and models and the connections between them
- analyse and explain how models and theories are used to organise and understand observed phenomena and concepts related to psychology, identifying limitations of selected models/theories

Develop

PART A

1 Study the terms in Table 9.10.

2 Study the concept map of the biological and psychological factors that affect visual perception in Figure 9.8.

3 Choose a term from Table 9.10 and write it in the correct space in Figure 9.8 to complete the concept map.

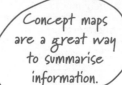
Concept maps are a great way to summarise information.

Table 9.10 Terms associated with visual perception

height in the visual field	texture gradient	linear perspective	proximity
binocular	convergence	shape	pictorial cues
orientation	memory	similarity	brightness
closure	figure ground	monocular	constancies

Figure 9.8 A concept map of the biological and psychological factors influencing visual perception

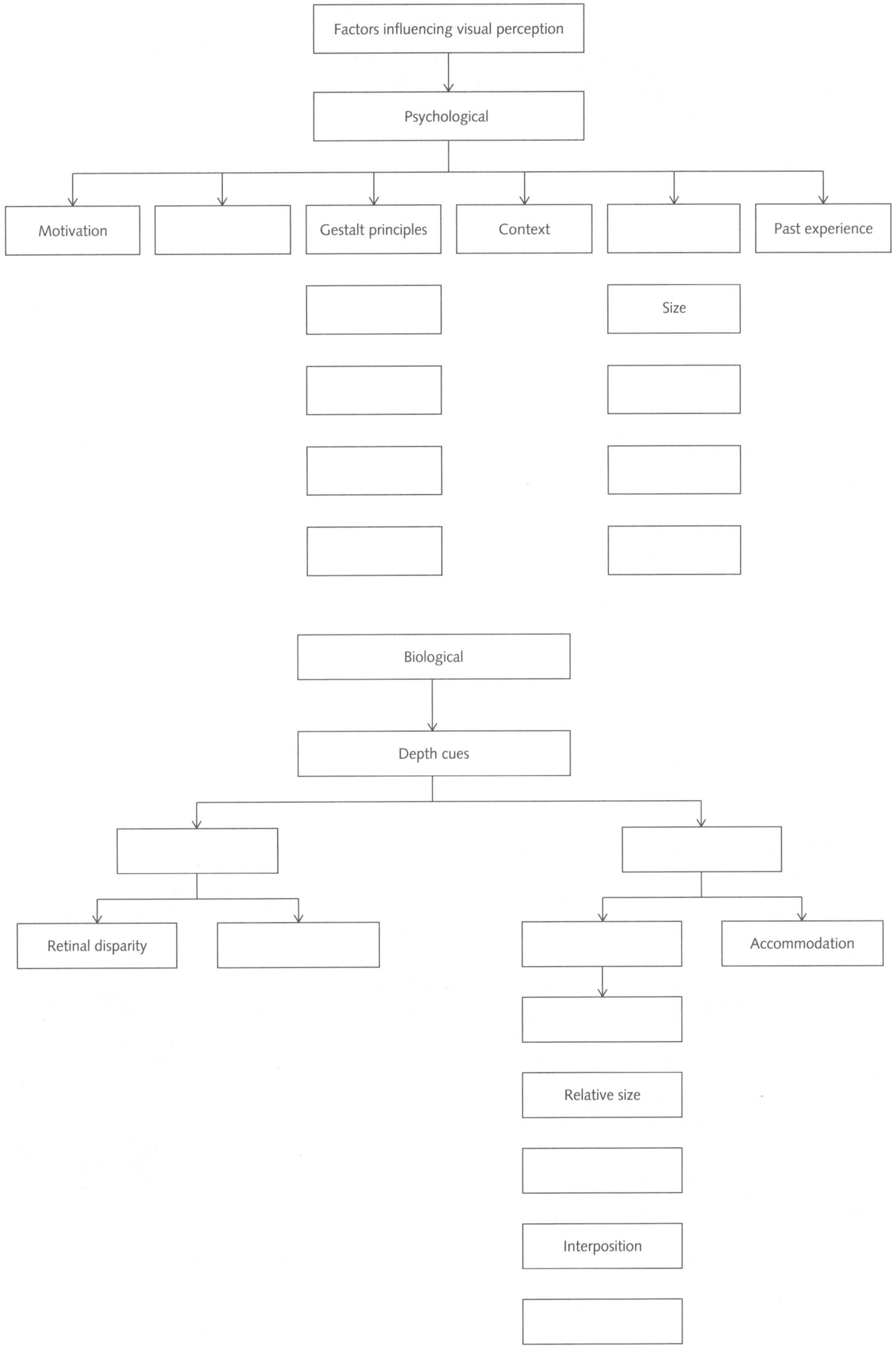

PART B

Visual constancies

Name the visual perception principle illustrated in each statement of the sections below.

1 When you place a round dinner plate on the table, you always perceive it as round regardless of the angle you view it from. _____

2 When you look at a yellow building and the sky becomes cloudy, you do not perceive the building as suddenly becoming a duller colour. _____

3 If you walk around your friend and then stand by their side, you do not perceive them as suddenly becoming thinner. _____

4 If you hang upside down on monkey bars, you still perceive the objects in your environment to be the right way up. _____

Name the depth cues illustrated in each statement

1 When you stand in front of a brick wall, you notice all the surface features of the wall, but when you view the wall from a distance, you do not notice these features. _____

2 If you look at a bus that is 20 metres away, your eyes receive two slightly different 2-D images of the bus but your brain combines the images so you perceive a single 3-D image of the bus. _____

3 If you look at the ocean and see boats just on the horizon, you perceive them to be further away than boats that are below the horizon. _____

4 If you look at the ocean and see ships sailing, when one ship crosses in front of another, the ship that is overlapped looks further away. _____

Name the gestalt principles illustrated in each statement

1 If you look at a clock on a striped wall, you would perceive the wallpaper to be the background, and the clock to be the object that is the centre of your focus. _____

2 When you look at a bowl of fruit, you group the fruits as bananas, oranges, apples, grapes and so on based on their comparable colour and shape. _____

3 An artist draws a few lines to suggest a well-known face and you can identify the person from these lines. _____

4 If three people stand near one another and a fourth person stands five metres away, the adjacent three will be seen as a group and the distant person as an outsider. _____

PART C

This activity will strengthen your ability to understand how pictorial depth cues work so that when you see a 2-D drawing or photograph, you are able to perceive it as 3-D.

1 Cut out Figure 9.9 and glue it onto the middle of a sheet of A3 paper.

2 Identify five pictorial depth cues in the photograph: linear perspective, texture gradient, interposition, height in the visual field and relative size.

3 Circle or highlight where each pictorial depth cue can be seen, and label it.

4 For each pictorial cue, draw an arrow to a blank piece of the poster and write answers to complete the following tasks:

- Define the pictorial depth cue.

- Explain how it is evident in the photograph.

- Explain how this depth cue helps you to perceive depth in the picture (that is, how does it help you to know which item is closer, what is the farthest point etc.?).

A cue may be evident in more than one place.

Figure 9.9 Identify and make a poster about the five pictorial depth cues in this photograph: linear perspective, texture gradient, interposition, height in the visual field and relative size

PART D

This activity will reinforce your understanding of the process of visual perception.

Use the terms in Table 9.11 to fill in the blanks in the concept map in Figure 9.10. It may be a good idea to enlarge Figure 9.10 onto A3 paper.

Table 9.11 Terms associated with the visual perception process

figure–ground	electromagnetic energy	stimulus	peripheral
closure	retinal disparity	pictorial cues	shape
context	transduction	convert	visual acuity
optic nerve	eye	movement	feature detector
organisation	understanding	psychological factors	reassemble
represents	electrochemical	orientation	retina
photoreceptors	ignore	meaningful	colour
			culture

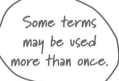

Some terms may be used more than once.

Figure 9.10 A concept map of the visual perception process

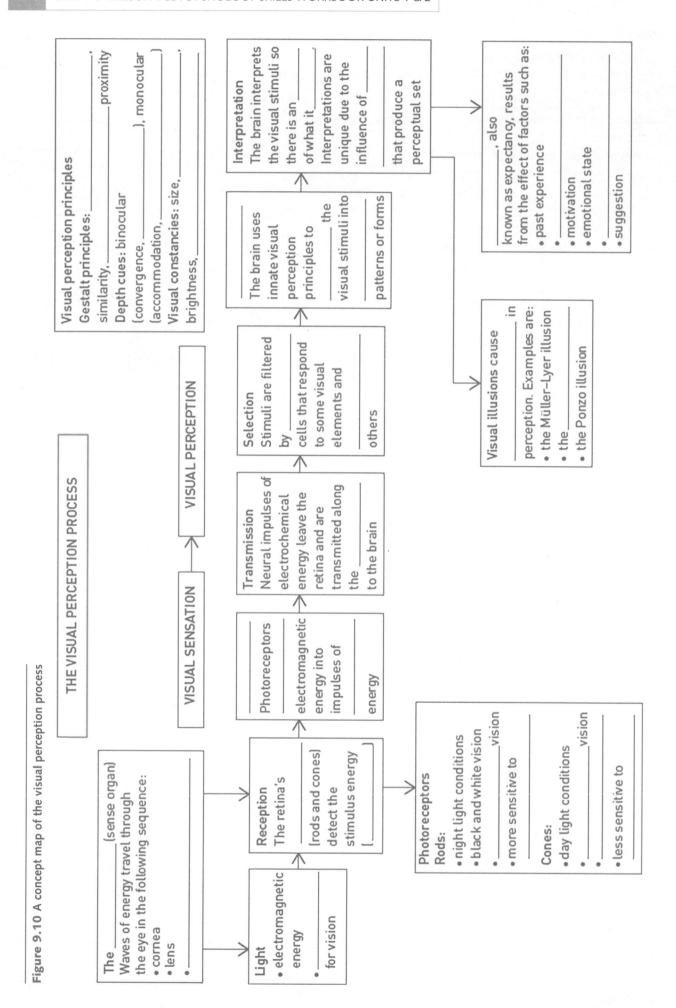

THE VISUAL PERCEPTION PROCESS

VISUAL PERCEPTION

VISUAL SENSATION

Visual perception principles
Gestalt principles:
similarity, _____, _____ proximity
Depth cues: binocular _____ (convergence, _____), monocular (accommodation, _____, _____, _____)
Visual constancies: size, _____, brightness, _____

Interpretation
The brain interprets the visual stimuli so there is an _____ of what it _____. Interpretations are unique due to the influence of _____ that produce a perceptual set

_____, also known as expectancy, results from the effect of factors such as:
• past experience
• _____
• motivation
• emotional state
• _____
• suggestion

Selection
Stimuli are filtered by _____ cells that respond to some visual elements and _____ others

The brain uses innate visual perception principles to _____ the visual stimuli into _____ patterns or forms

Visual illusions cause _____ in perception. Examples are:
• the Müller–Lyer illusion
• the _____
• the Ponzo illusion

Transmission
Neural impulses of electrochemical energy leave the retina and are transmitted along the _____ to the brain

Reception
The retina's _____ (rods and cones) detect the stimulus energy [_____]

Photoreceptors _____ electromagnetic energy into impulses of _____ energy

Light
• electromagnetic energy
• _____ for vision

The _____ (sense organ)
Waves of energy travel through the eye in the following sequence:
• cornea
• lens
• _____

Photoreceptors
Rods:
• night light conditions
• black and white vision
• _____ vision
• more sensitive to _____

Cones:
• day light conditions
• _____ vision
• _____
• less sensitive to _____

9.4   Gustatory perception

9.4.1 Sensation and perception of taste

Develop

This activity will reinforce your understanding of structures and concepts associated with the sense of gustation. The wordfind puzzle in Figure 9.11 contains 11 terms associated with the gustatory system. The clues for these are given in the statements below. Find the answers in the wordfind.

1   The portion of the individual gustatory receptor's surface, which receives the stimulus that triggers the neuron to fire. _____

2   The minimal amount of stimulus energy an individual requires to be conscious of a sensation. _____

3   The sensory experience of a food or drink that is perceived as flavour. _____

4   A taste sensation produced by glutamate and several amino acids that is experienced as meaty or savoury. _____

5   The type of nerves that receive sensory information from the taste buds and transmit it to the medulla. _____

6   Tiny structures located on the papillae on the tongue's surface.

_____

7   The small bumps on the tongue's surface. _____

8   The receptor cells for taste; located on the surface of the taste buds. _____

_____

9   The brain structure that receives all sensory input and transmits it to the appropriate brain area for further processing. _____

10 Related to the way a food or beverage feels in the mouth. _____

11 A brain stem structure that receives the sensory information related to a food or beverage from the cranial nerves. _____

Remember that the answers can be horizontal, vertical or diagonal, and forwards or backwards.

Figure 9.11 Gustation word find

D	T	E	E	M	E	D	U	L	L	A	Q	S	S	N	N	U	M	I	F
R	E	C	E	P	T	I	V	E	F	I	E	L	D	I	Y	P	Y	Y	R
E	F	N	A	A	H	H	B	I	R	U	R	E	E	F	R	A	R	T	I
C	T	G	G	B	A	D	F	C	F	H	M	O	O	A	F	P	R	H	D
Q	T	U	S	U	S	O	H	C	L	A	B	Y	G	L	L	I	P	A	P
X	S	S	S	M	R	O	F	N	R	H	A	G	A	Q	G	L	G	L	P
O	C	T	H	A	L	E	L	I	D	L	U	A	O	N	H	L	T	A	O
B	H	A	E	M	A	N	S	U	R	S	M	M	B	B	C	A	T	M	O
E	T	T	H	X	C	A	T	Y	T	K	G	R	A	A	R	E	L	U	K
E	H	O	T	U	T	S	P	A	A	E	L	L	Y	R	K	U	U	S	N
G	A	R	G	D	D	E	T	U	S	H	T	G	E	T	T	I	Y	C	L
L	L	Y	H	L	Y	I	A	T	T	S	H	H	Z	O	O	O	H	C	C
Y	L	R	E	C	O	R	S	S	E	A	E	U	R	S	T	A	S	R	E
D	A	E	G	N	L	E	N	H	B	C	B	C	F	E	E	Y	F	A	L
V	A	C	Y	Y	R	P	O	E	U	N	E	H	X	O	S	F	Y	N	N
O	B	E	G	B	R	I	Y	F	D	I	E	T	B	Z	L	H	L	I	B
I	U	P	A	P	I	Z	O	B	S	O	U	S	P	N	O	R	O	A	C
D	S	T	A	S	T	Y	H	A	I	R	O	O	M	C	R	R	H	L	B
Y	S	O	N	I	M	Z	W	E	E	B	W	W	L	L	O	O	G	C	D
E	B	R	E	C	C	E	P	C	O	Z	U	U	M	A	M	I	N	C	P
K	M	S	O	T	B	J	N	R	A	W	U	X	W	Q	Y	U	I	P	C

9.4.2 Applying the biopsychosocial model to gustatory perception

Key science skills

Analyse, evaluate and communicate scientific ideas

- discuss relevant psychological information, ideas, concepts, theories and models and the connections between them
- analyse and explain how models and theories are used to organise and understand observed phenomena and concepts related to psychology, identifying limitations of selected models/theories

Develop

Categorise the factors in Table 9.12 as biological, psychological or social factors by writing them in the correct area in Figure 9.12.

Table 9.12 Factors influencing taste perception

upbringing	disease	culture
context	texture	level of education
time between meals	mood	description/language
learning	smell	pregnancy
age	past experience	hunger
utensils	environment	expectations
colour	genetics	socio-economic status
temperature of food	arrangement of food	peers
early exposure	gut bacteria	appearance
occupation	memory	moisture content
hormones	attention	minimum threshold
packaging	travel	cost
light levels	plate colour	advertising
beliefs and values	religion	motivation

Figure 9.12 Applying the biopsychosocial model

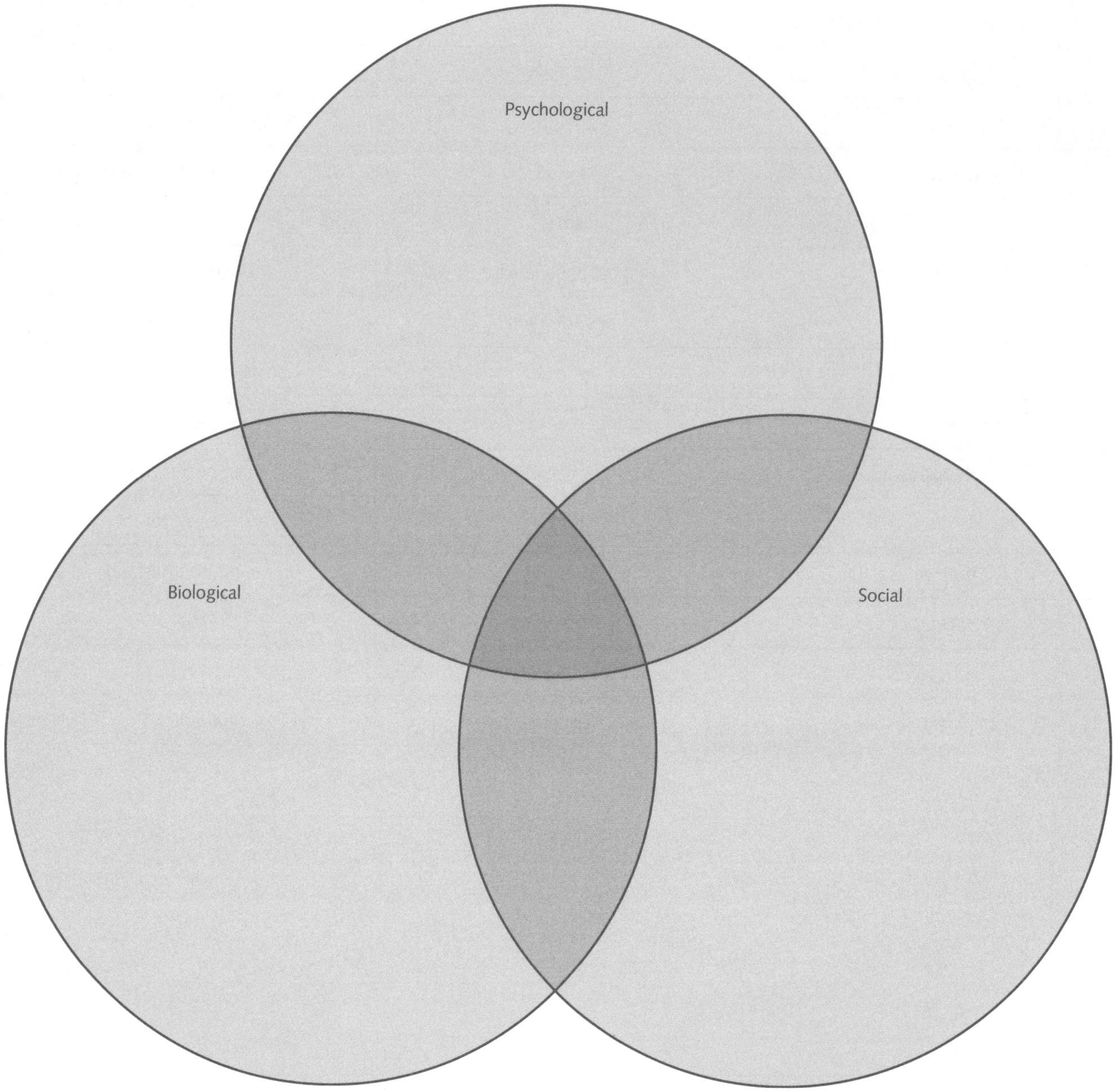

Chapter 9 summary

1　Read the statements in the table below.

2　Place a tick in the correct column to indicate whether each statement is true or false.

Table 9.13 True or false?

Statement	True	False
Reception involves sensory receptors converting stimulus energy into impulses of electrochemical energy.		
Sensation is an automatic physical process.		
Perceptions of a common stimulus are the same for all individuals.		
Selective attention allows us to 'tune out' irrelevant information so we can focus on important information.		
Sustained attention peaks during the early 40s.		
Perception always starts with the processing of low-level features (tiny details) of sensory stimuli.		
Automatic processing requires a high level of attention and monitoring.		
Multitasking often does not work well because our attention is limited.		
Sensory stimuli selected for attention receive further processing by the brain.		
Depth cues are a psychological factor that influences visual perception.		
Pictorial cues require binocular vision.		
Figure–ground involves the viewer using an imaginary contour line to perceptually group and separate some features of a stimulus against a background.		
Most perceptions are unaffected by learning.		
An object's perceived level of brightness relative to its surroundings stays the same under changing light conditions.		
Culture does not cause differences in visual processing between individuals.		
There are five primary tastes that humans are sensitive to.		
Cones are more sensitive and numerous than rods.		
Taste is not dependent on smell.		
Taste buds are only found on the tongue.		
Taste buds contain rods and cones.		
Genetics can influence our perception of taste.		
Taste perception is influenced before birth.		
Adults have more taste buds than children.		
The amygdala is involved in developing taste aversion.		
All stimuli processed during sensation are processed during perception.		

10 Distortions of perception

10.1 Fallibility of visual perception systems

Key knowledge
- the fallibility of visual perceptual systems, for example visual illusions and visual agnosia

10.1.1 Visual illusions and visual agnosia

Key science skills
Analyse, evaluate and communicate scientific ideas
- discuss relevant psychological information, ideas, concepts, theories and models and the connections between them

Develop

Use the terms in Table 10.1 to fill in the blanks in the following paragraphs.

Table 10.1 Terms used to describe visual illusions and visual agnosia

distort	cues	error
all	mismatch	false
principles	consistent	reality
sensory	identify	single
integrated	rare	stored

A visual illusion is a _____ perceptual _____ in interpreting the features of an external stimulus. Illusions _____ stimuli that actually exist and lead us to misapply _____ we use to organise stimuli into a stable, consistent or meaningful perception. This causes us to make a _____ judgement of reality.

When we experience a visual illusion, the stimulus provides us with _____ that mislead our perception. As a result, a _____ between our perception and the _____ of the actual stimulus occurs. Although perception of common stimuli is often described as unique, this is not the case with a visual illusion. With visual illusions, _____ individuals tend to be misled by the stimulus in the same way. In a visual illusion, length, position, motion, curvature or direction of the stimulus is consistently misjudged.

9780170465045

Agnosia is a _____ condition caused by a neurological disorder or damage to parts of the brain. It results in an inability to process _____ information. People with agnosia fail to recognise and _____ objects, people, smells or sounds despite otherwise normally functioning senses. Agnosia only affects a _____ sensory processing system, for example, vision.

Damage has occurred in pathways that connect their brain's _____ lobe with those areas involved in visual memory. If they are damaged, the _____ information cannot be accessed. Therefore, it cannot be _____ with the sensory information received.

10.1.2 Who am I?

Key science skills
Analyse, evaluate and communicate scientific ideas
- discuss relevant psychological information, ideas, concepts, theories and models and the connections between them

Develop

Use the terms in Table 10.2 to complete the following 'Who am I?' questions. Not all terms match a question.

Table 10.2 Terms associated with the fallibility of visual perception

agnosia	associative visual agnosia	spinning dancer illusion	circular	Ponzo illusion
Müller-Lyer illusion	apperceptive visual agnosia	illusion	prosopagnosia	Zollner illusion
Kanizsa triangle	perceptual expectancy	Ponzi illusion	binocular cues	visual perception principles

1   When people look at my two lines, they perceive the line that has arrows pointing inwards on its ends to be shorter than the line that has arrows pointing outwards. However, in reality, my two lines are of equal length. _____

2   I am unable to recognise visual stimuli, such as shapes or forms of an object, despite having no visual deficits. _____

3   When people look at the Ames room through a peephole, they misperceive the size of the people inside because they cannot use me to gain enough information to make a correct perception. _____

4   You see me as a white triangle, but I don't actually exist! _____

5   My lines look crooked but they aren't! _____

6   I am unable to recall information associated with an object, such as its name or what it is used for. _____

7   I am a rare condition caused by a neurological disorder or damage to parts of the brain. It results in an inability to process sensory information. _____

8 I am a consistent misperception of a real stimulus. _____

9 I am an illusion where two horizontal lines of equal length are drawn inside two converging lines, but the line in the narrower section of the converging lines is perceived to be longer. _____

10 I am unable to recognise familiar faces, sometimes even my own! _____

11 My image appears to spontaneously switch direction between left and right. _____

12 If you live in my type of environment, you tend not to be affected by the Müller-Lyer illusion. _____

13 I rely on factors such as past experience and context to create a readiness to perceive visual stimuli in a particular way. _____

14 I am the rules the brain applies to construct perceptions. _____

10.2 Fallibility of gustatory perception

Key knowledge
- the fallibility of gustatory perception, for example supertasters, exposure to miraculin and the judgement of flavours

10.2.1 Are you a supertaster?

Key science skills

Develop aims and questions, formulate hypotheses and make predictions
- identify, research and construct aims and questions for investigation

Generate, collate and record data
- systematically generate and record primary data, and collate secondary data, appropriate to the investigation
- organise and present data in useful and meaningful ways, including tables, bar charts and line graphs

Analyse and evaluate data and investigation methods
- process quantitative data using appropriate mathematical relationships and units, including calculations of percentages, percentage change and measures of central tendencies (mean, median, mode), and demonstrate an understanding of standard deviation as a measure of variability
- identify and analyse experimental data qualitatively, applying where appropriate concepts of: accuracy, precision, repeatability, reproducibility and validity; errors; and certainty in data, including effects of sample size on the quality of data obtained

Develop

Taste buds, located on small bumps on the tongue called fungiform papillae, are each made up of about 50 to 150 taste receptor cells. On the surface of these cells are receptors that bind to small molecules related to flavour. Each receptor is best at sensing a single flavour: sweet, salty, bitter, sour or umami. The sum total of these sensations is the 'taste' of the food.

The number of taste buds varies from person to person. People who have relatively more taste buds are called supertasters. Non-tasters have very few taste buds and, to them, most food may seem bland and unexciting. The people in the middle are average tasters.

A student conducted an experiment to measure the average number of taste buds found on the tongues of different individuals. As part of the experiment, the student took images of different 1 cm² sections of the tongues of different participants as shown below.

Images of different 1 cm² sections of each participant's tongue		
Section 1	Section 2	Section 3

Person 1, Person 2, Person 3 — taste bud (labelled)

PART A

1   Create a table to record the data in the space provided. Include in your table the average for each participant. Remember to review the principles of constructing data tables as outlined in your textbook.

Table 10.3 Mean number of tongue papillae for each participant

2   Other students in the class also conducted the same experiment. Below are the results obtained by other students in the experiment.

Using both yours and the class average papillae data, calculate the:

- mode

_____

- median.

_____

Participant	Average number of papillae/cm²
A	22
B	13
C	24
D	30
E	31
F	21

3   Construct a frequency histogram of the combined average data.

_____

Figure 10.1 Frequency of tongue papillae in a sample, n= _____ [insert the total sample number]

4   What was the aim of this investigation?

_____

_____

5   Describe the frequency histogram.

_____

_____

_____

Figure 10.2 Density of tastebuds in supertasters (n=10), medium tasters (n=15), and non-tasters (n=14).

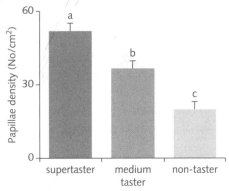

Source: Melis, M., Sollai, G., Mastinu, M., Pani, D., Cosseddu, P., Bonfiglio, A., Crnjar, R., et al. (2020). Electrophysiological responses from the human tongue to the six taste qualities and their relationships with PROP taster status. *Nutrients*, 12(7), 2017. MDPI AG. Retrieved from http://dx.doi.org/10.3390/nu12072017

Figure 10.2 shows the results of an experiment involving the density of tastebuds in different types of tasters. Refer to the data in Figure 10.2 and the combined class data to answer the following questions.

6 Based on Figure 10.2, what range of papillae density would someone have to be considered a supertaster?

7 Using the combined data, what percentage of participants are considered supertasters?

8 Based on the combined data, what percentage are non-tasters, having fewer than 15 papillae?

9 What percentage have between 15 and 30 papillae, and are average tasters?

10 Define accuracy.

11 How accurate were the results originally collected by the student? Justify.

PART B

Some variations to try include the following.

1 Use an equal number of male and females, with at least 15 of each. Do males and females have the same likelihood of being a supertaster?

2 Look into how different foods, like spinach and chilli, taste to the different types of tasters. Based on this, can you make a taste test to figure out who is a supertaster? Check how accurate your taste test is by also counting the number of papillae for each person.

3 Use a person's body mass index (BMI, formulas can be found online) and their results to try to figure out if there is a correlation between a person's weight and the type of taster they are (supertaster, non-taster, or average taster).

4 Conduct a test with participants from the same family to see if there is support for there being a genetic link.

Here are some variations to try!

Source: activity adapted from Science Buddies Staff (2020, November 20). Do you love the taste of food? Find out if you are a supertaster! https://www.sciencebuddies.org/science-fair-projects/project-ideas/HumBio_p017/human-biology-health/find-out-if-you-are-a-supertaster

10.2.2 Judging flavours (part 1)

Key science skills

Develop aims and questions, formulate hypotheses and make predictions

- identify, research and construct aims and questions for investigation
- identify independent, dependent, and controlled variables in controlled experiments
- formulate hypotheses to focus investigation

Generate, collate and record data

- systematically generate and record primary data, and collate secondary data, appropriate to the investigation

Develop

IMPORTANT NOTE

As with any experiment/activity that calls for students to ingest a product, please be sure to get permission from the students' parent(s)/guardian(s) prior to conducting the activity and/or administration approval before you engage this activity.

Make sure to check for any allergies.

MATERIALS

- blindfold
- 1 cup containing six jellybeans (3 different coloured pairs)
- 1 cup of water

INSTRUCTIONS

1 Using the label from the jellybean packaging, write down the colour and corresponding flavour of the three sets of jellybeans in column 1 of Table 10.4.
2 Give the participant one of the jellybeans to try. The participant must have their eyes open so they can see the colour of the jellybean.
3 Ask the participant what the perceived flavour of the jellybean is and write down the response in column 2. The participant must then take a small drink of water.
4 Repeat this for the two other colours, asking them to drink a small amount of water after eating each jellybean.
5 Blindfold the participant and repeat the above step (participants must still drink water after each jellybean), recording each response in column 3. The jellybeans should be given out in a different order to before (e.g. if the first time was yellow/lemon, red/raspberry, pink/strawberry, then change to something like red/raspberry, pink/strawberry, yellow/lemon). Do not tell the participant which colour jellybean they are receiving.

Table 10.4 Judging the flavour of jellybeans

Colour and flavour of each set of jellybeans (according to packet)	Perceived flavour of jellybean, eyes open	Perceived flavour of jellybean, eyes closed

1 What is the aim for this experiment?

2 What is the independent variable (IV)?

3 What is the dependent variable (DV)?

4 What was the research design?

5 What are the advantages of this type of design?

6 What are the limitations of this type of design?

7 Suggest a method that addresses these limitations.

8 Write a research hypothesis for the experiment.

10.2.3 Judging flavours (part 2)

> **Key science skills**
>
> Comply with safety and ethical guidelines
> - demonstrate safe laboratory practices when planning and conducting investigations by using risk assessments that are informed by safety data sheets (SDS), and accounting for risks
>
> Generate, collate and record data
> - systematically generate and record primary data, and collate secondary data, appropriate to the investigation

Develop

Miraculin is a protein that binds to the sweet taste receptors on the tongue, causing sour and bitter foods to taste sweet. Scientists have yet to discover why miraculin appears to have this effect on certain taste receptors, but some hypothesise that miraculin briefly alters the shape of the sweet taste receptors, making them responsive to bitter and sour tastes.

IMPORTANT NOTE

As with any experiment/activity that calls for students to ingest a product, please be sure to get permission from the students' parent(s)/guardian(s) prior to conducting the activity and/or administration approval before you engage this activity.

There do not appear to be any known side effects from this tablet/berry. Since the tablet/berry does not change any chemical compound of any of the foods that are used, they will retain their high acidity. Be careful not to ingest high quantities of highly acidic foods as this can lead to oral ulcers (due to high levels of acidity).

MATERIALS

- Miracle Fruit tablets, available from Questacon Shop (https://shop.questacon.edu.au) or other online retailers. One tablet per person; the larger tablets can be split in half.
- Food items cut into small wedges with seeds removed for easy consumption and placed into tasting cups. Examples include lemons, limes, grapefruit, kiwi, cranberries, pineapple, green grapes, green mangoes, Granny Smith apples, pickles, and apple cider vinegar (use in very small doses).

INSTRUCTIONS

1 Have students taste some of the sour/bitter foods prior to the activity in case they are unsure of their taste.

2 Ask them to rinse their mouth out with plenty of water to clean their palate.

3 Ask students to put one (or one half) tablet on their tongue and let the substance dissolve (this should take approximately 3 minutes).

4 Once the tablet is fully dissolved, students can pick up a wedge of the sour/bitter food and proceed to put in their mouths and eat it.

Source: activity adapted from Neuroscience for Kids. (n.d.). Can it really be that sweet? https://faculty.washington.edu/chudler/sweet.html

RESULTS

Table 10.5 Rating scale between 1 to 5 of perceived sweetness with 1 = neutral, 3 = similar to table sugar, 5 = extremely sweet, with 2 and 4 a rating between these descriptions

Food item	Rating of sweetness (1–5)	Rating (class average) (1–5)

1 Based on the results, what impact does the Miracle Fruit tablet on the taste of bitter/sour fruits?

2 What type of data has been collected?

3 What are the advantages of a self-report measure?

4 What are the disadvantages of a self-report measure?

10.3   Distortions of perception in healthy individuals

Key knowledge
- distortions of perception of taste and vision in healthy individuals such as synaesthesia and spatial neglect

10.3.1 Distortions of perception: visual

Key science skills
Analyse, evaluate and communicate scientific ideas
- discuss relevant psychological information, ideas, concepts, theories and models and the connections between them

Develop

Complete the overview of distortions of perception in Figure 10.3 by providing the required information.

Figure 10.3 Concepts related to distortions of perception

Perceptual fallibility occurs when:

A perceptual distortion occurs when:

A hallucination is:

A visual illusion is:

A perceptual anomaly is:

A perceptual disorder is:

Synaesthesia is:

It is a perceptual anomaly, not a perceptual disorder, because:

10.3.2 Evaluation of research

Key science skills

Construct evidence-based arguments and draw conclusions

- use reasoning to construct scientific arguments, and to draw and justify conclusions consistent with evidence base and relevant to the question under investigation

Develop

Read the information about synaesthesia and complete the questions that follow.

Research suggests that synaesthesia may help people express their creativity and that synaesthetes may be more common in the world of creative art. For example, in the general population a reliable estimate of the prevalence of grapheme–colour synaesthesia (when specific numbers, letters or words trigger specific colours) is about one per cent. However, it has been claimed that synaesthesia is more common in creative artists.

In 2010, two researchers from the University of Bern in Switzerland, Rothen and Meier, conducted a study to test whether there really is a higher prevalence of grapheme–colour synaesthesia in artists. The researchers tested a volunteer group comprised of students from the art college of Zurich, Switzerland, and visitors to an open-house event at the University of Bern. All participants were tested with a computerised grapheme–colour consistency test. They were individually presented with 36 graphemes, one at a time, in random order. Each grapheme was accompanied, on the same screen, by a palette of 13 basic colours, the same each time but randomly arranged on each trial. Participants were required to select the 'best' colour for each grapheme. After an initial presentation, an immediate surprise retest followed, in which the graphemes were presented again in a re-randomised order. The consistency score was calculated as the number of identical grapheme–colour associations. Afterwards, participants responded to six statements concerning their experiences. For example:

'Whenever I see or think about letters or numbers (printed black-on-white), I automatically experience the letter or number as having another colour (e.g. red).' A value of between 0 and 5 was assigned to each statement, where high scores reflect more typical synaesthetic responses. Synaesthetes were identified by their results on both consistency and questionnaire scores.

In the sample of art students, seven participants were classified as synaesthetes. In the control sample, however, only two participants were classified as synaesthetes. The proportion of synaesthetes was significantly higher for the art students than in the control sample. Furthermore, the consistency scores were higher for the art students than for the control samples. Additionally, the questionnaire scores were also higher for the art students than for the control sample. The results indicate a higher prevalence of synaesthesia in the sample of art students than in the general population.

The researchers suggested that it is possible that the higher prevalence of grapheme–colour synaesthetes in art students is due to the richer world of experiences provided by the synaesthetic associations, and hence their skill and choice of interest in art as a hobby or career. Furthermore, synaesthetic predispositions may become activated through specific exercises and practices in art education.

Source: Higher prevalence of synaesthesia in art students by Nicolas Rothen and Beat Meier, *Perception*, 2010, volume 39, pages 718–720, SAGE Publications, doi:10.1068/p6680

9780170465045

1 What was the aim of this study?

_____

_____

2 Write a hypothesis for this study.

_____

_____

3 Identify the experimental group and the control group in this study.

- Experimental group

_____

- Control group

_____

_____

4 Identify the research design used in this study and state one advantage and one limitation of this design.

- Research design

_____

- Advantage

_____

_____

- Limitation

_____

_____

5 Identify the sampling method used in this study.

_____

6 State the results of this study.

_____

_____

_____

7 What conclusion(s) can be drawn from these results?

_____

_____

_____

_____

8 Identify and define two ethical concepts that the researchers would have had to follow in the conduct of this study.

_____

_____

_____

_____

_____

_____

10.3.3 Synaesthesia

Key science skills

Analyse, evaluate and communicate scientific ideas

- discuss relevant psychological information, ideas, concepts, theories and models and the connections between them

Develop

PART A

Indicate whether each statement is true or false by placing a tick in the correct column in Table 10.6.

Table 10.6 Synaesthesia: true or false statements

Statement	True	False
When people with synaesthesia are exposed to a specific stimulus, the stimulus triggers activity in two other senses that would not be triggered in people who do not have synaesthesia.		
With specialist training, most people could develop synaesthesia.		
Neuroscientists now accept that synaesthesia probably has a biological base and, therefore, is a genuine neurological condition.		
If you automatically think of the colour red or the sound of Christmas carols when you hear the word 'Christmas', you probably have synaesthesia.		
One theory suggests that synaesthesia is the result of reduced apoptosis (the pruning of weak, excess or unused neurons) in infancy, which leads to increased sensory connectivity between brain areas.		
One person can have a number of subtypes of synaesthesia.		
If a person with synaesthesia 'sees' the colour red every time they hear the number 10, when they see the colour red they will also 'hear' the number 10.		
Studies suggest that there is a positive correlation between synaesthesia and creativity.		
Most synaesthetes experience a reduction in their synaesthesia as they grow older.		
Synaesthesia is classified as a perceptual disorder.		

PART B

1   Study Figure 10.4.
2   In the 'Synaesthesia' circle, write a definition for synaesthesia.
3   Read the following list of bullet points:
   - It appears to have a genetic link.
   - Grapheme–colour synaesthesia
   - The trigger always causes the same synaesthetic experience.
   - It requires a specific trigger.
   - Synaesthetes are born with extra neural connections that cross over from one area of the cerebral cortex to another area.
   - Sound-to-colour synaesthesia
   - The experience is always one-directional.
   - Lexical–gustatory synaesthesia
   - It is more common among creative people.
   - There is failure to prune excess neurons during infancy.
   - Personification
   - It is an involuntary response to a stimulus.
   - Number–form synaesthesia
   - The experience is unique for the individual.
   - Synaesthesia may be linked to learning during childhood when children learn to associate numbers or letters with colours, possibly to aid memory.
4   Write each bullet point under the correct heading in the rectangles ('Common characteristics', 'Possible causes' and 'Common types'). Hint: The number of bullet points appropriate for each rectangle has been indicated.

Figure 10.4 Summary of synaesthesia

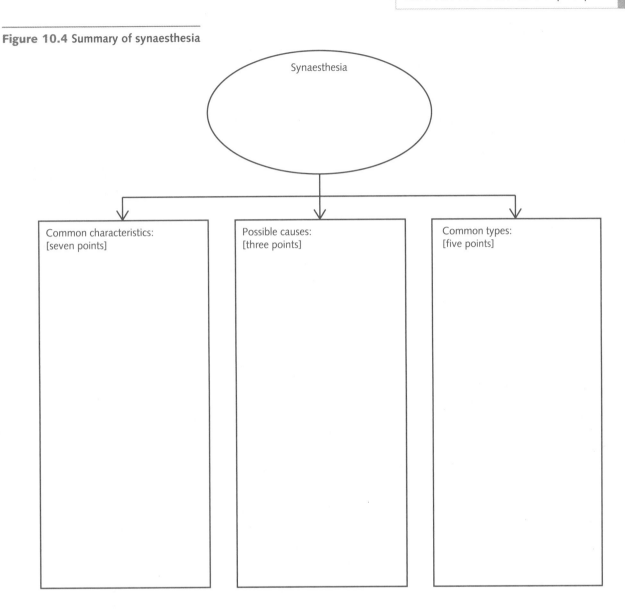

Synaesthesia

Common characteristics:
[seven points]

Possible causes:
[three points]

Common types:
[five points]

10.3.4 Spatial neglect

Key science skills
Analyse, evaluate and communicate scientific ideas
- discuss relevant psychological information, ideas, concepts, theories and models and the connections between them

Develop

Spatial neglect is a tendency to ignore the left or right side of one's body or the left or right side of visual space. It is not a result of visual impairments. Primarily, it is a cognitive impairment because it is a disorder of attention.

PART A

MATERIALS

- eyeglasses
- masking tape/paper

INSTRUCTIONS

1 Wear a pair of eyeglasses that have half of both lenses taped or blacked out on the same side of each lens, taking care to remove any peripheral vision.
2 Under supervision complete a series of everyday activities for 2 minutes.

3   Describe your experience.

- Behaviours

_____

_____

_____

_____

- Cognitions

_____

_____

_____

_____

- Emotions

_____

_____

_____

_____

> This adds to the work covered in Chapter 5.

PART B

Occupational therapists use several strategies to assist people with spatial neglect. Examples are of these strategies are provided.

- Encourage the use of the affected side. For example, using hand-over-hand guiding to help the patient use their fork, wash their face and body with the affected hand, squeeze toothpaste or brush their hair.
- Anchoring. Instruct the patient to scan to the affected side by providing visual cues, like a bright sticky note or coloured tape, and train the patient to scan looking all the way to the anchor and back (like a light from a lighthouse). Always sit at the patient's affected side, as this can help to cue the individual to look more towards their affected side.
- Visual scanning. Incorporate visual scanning activities like letter cancellation worksheets or wordfinds.
- Use a mirror. This provides visual feedback to the patient and can aid in mid-line orientation as the person can help to correct their leaning posture.

Source: adapted from Stromsdorfer, S. (n.d.). Occupational therapy interventions for unilateral neglect. *My OT Spot*.

For each example in Table 10.7, suggest a strategy that can be used by family, in the home, to assist someone with spatial neglect. Note: assume the damage has occurred to the right side of the brain.

Table 10.7 Spatial neglect affects many everyday activities

Activity	Strategy
Colliding with obstacles on their left as they walk from one room to another	
Eating food only from the left side of their plate	
Ignoring people situated on their left	
Failing to shave the left side of their face	
Back pain due to an uneven posture	

PART C

Complete the following cancellation worksheets.

1 Cross out all of the letter 'H' from this series of letters in Figure 10.5.

Figure 10.5 A cancellation activity with the letter H.

Cancellation sheet

B H D F C H C F H G I H C H I H B D A H C F B H D E H D A F H I C H F H B A F H E H F H C B D H F G H E
H E G H F E H D H F H C B F H A D H C E H I H G D H G E B H E G H I H C H E H F C I H E B H G F D H B E
H B H A E H B H C F A H F H G H C G D H C B A H G D E H C H B E H D G H D A F H B I F H E B H D H E H G
H D G A H C H F B H A F H E B F H C D H F H G E H B H D H F A C H C H F D I H C B I H B H A C H D H F B
E H B H G B I H C E H A F H I H E B H G F B H F A H E B G H G F E H D B H B H C F H A D C H E I H F H G
H D C B H E D G H A D F H B H I G E H G H D E H C G H D H E B A H F B H C D A H G B H C H D F H C A I H

Your score is calculating by subtracting the number of H's that were not crossed out from the perfect score of 104 (0-53 on the left and 0-51 on the right).

A higher score indicates better performance.

The presence of spatial neglect can be inferred by calculating the frequency of errors to the left or to the right from the middle of the page.

2 Cross out all the small stars in Figure 10.6.

Figure 10.6 A cancellation activity involving stars.

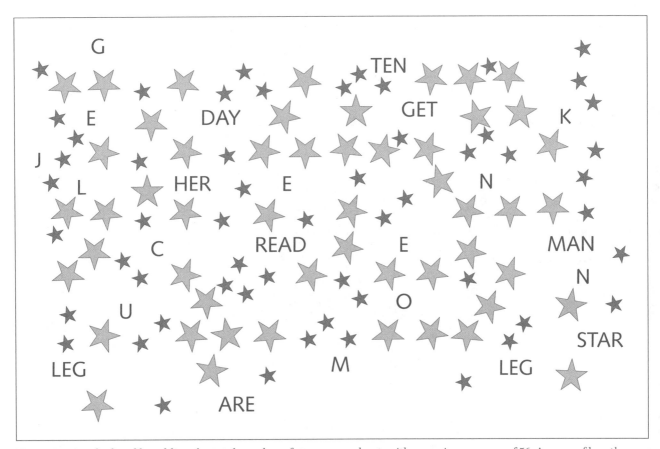

Your score is calculated by adding the total number of stars crossed out, with a maximum score of 56. A score of less than 44 indicates the presence of spatial neglect.

A Laterality Index or Star Ratio can also be calculated from the ratio of stars crossed off on the left of the page to the total number of stars cancelled. Scores between 0 and 0.46 indicated USN in the left hemispace and scores between 0.54 and 1 indicate USN in the right hemispace.

Chapter 10 summary

PART A

The word find puzzle in Figure 10.7 contains 15 terms associated with how visual and taste sensations can be distorted. The clues for these are given in the numbered statements that follow. Find the answers in the wordfind. (Note that the answers can be horizontal, vertical or diagonal, and forwards or backwards.)

1 A perceptual error resulting in the perception of an object or event that has no external reality. _____

2 An irregularity in perception that causes the experience of sensations that are not consistent with a stimulus. _____

3 A consistent perceptual error in interpreting a real external stimulus. _____

4 The hypothesis that states that an apparently more-distant object will be perceived as larger than an apparently nearer object with the same-sized retinal image. _____

5 This type of stimulus can be interpreted in more than one way. _____

6 Information provided by a food that we use when forming a perception of the food. _____

7 Information connected to a food, but not part of the food, that we use when forming a perception of the food. _____

8 We tend to notice this first about a food. _____

9 The term used to describe the synaesthetic characteristic whereby a specific colour might trigger the perception of a specific taste, but exposure to the taste doesn't trigger the perception of the colour. _____

10 A perceptual anomaly where activation of one sense triggers the activation of another sense not normally activated by the same stimulus. _____

11 Because people are born with synaesthesia and they cannot control it, it is described as this. _____

12 Most researchers agree that a person with synaesthesia has more of these. _____

13 Synaesthesia is more common in this group of people. _____

14 The person accredited with first recording cases of synaesthesia. _____

15 The term to describe people with synaesthesia. _____

Figure 10.7 Summary word find

A	S	E	E	E	A	R	Y	U	C	P	P	G	F	G	W	R	R	S	A
W	U	N	E	U	R	A	L	C	O	N	N	E	C	T	I	O	N	S	R
Y	R	F	K	K	T	J	Z	C	L	U	X	X	V	G	C	C	O	S	I
U	I	G	K	K	I	J	F	R	O	F	V	T	V	G	A	Y	I	Y	N
U	L	X	C	Z	S	J	S	R	U	R	R	G	E	W	Y	T	Y	T	
P	L	O	U	R	T	E	Y	U	R	E	E	I	R	V	R	R	A	P	R
F	U	U	Y	G	I	U	N	A	Y	O	O	N	O	O	B	C	N	Z	I
Z	S	A	A	C	C	R	A	K	J	O	K	S	W	V	B	C	I	Z	N
Z	I	K	Z	O	G	Z	E	T	A	E	T	I	G	K	X	B	C	K	S
F	O	F	C	V	O	G	S	P	U	M	P	C	G	R	V	X	U	D	I
K	N	E	C	K	W	A	T	W	A	W	B	B	T	D	U	U	L	U	C
D	D	A	S	P	P	S	H	I	L	L	O	I	B	J	D	P	L	I	S
G	G	F	Y	A	S	D	E	O	K	P	D	O	G	Z	X	J	A	E	S
D	A	G	V	F	K	V	S	J	A	S	E	D	O	U	B	N	H	U	Y
T	J	L	Y	E	X	J	I	O	X	E	T	T	I	I	O	E	U	S	N
E	K	K	T	V	V	X	A	R	T	L	T	B	R	M	E	U	X	W	A
Y	C	C	F	O	B	B	P	P	G	X	T	K	A	A	U	L	S	S	E
D	Z	Z	J	X	N	T	J	C	C	X	L	L	P	D	K	I	I	U	S
T	I	N	V	O	L	U	N	T	A	R	Y	Y	D	D	A	S	W	E	T
I	U	E	A	A	W	E	S	R	W	A	S	E	R	R	I	O	N	D	H
A	P	P	A	R	E	N	T	D	I	S	T	A	N	C	E	Y	L	U	E
Y	I	O	N	E	D	I	R	E	C	T	I	O	N	A	L	T	T	R	T
I	P	D	J	P	T	A	E	T	P	D	U	D	D	F	F	J	P	P	E
S	A	S	E	T	L	L	K	K	F	G	G	J	T	A	J	J	R	W	S

PART B

The crossword puzzle in Figure 10.8 contains terms associated with distortions of perception. Use the clues to complete the crossword.

Figure 10.8 Distortion of perception crossword

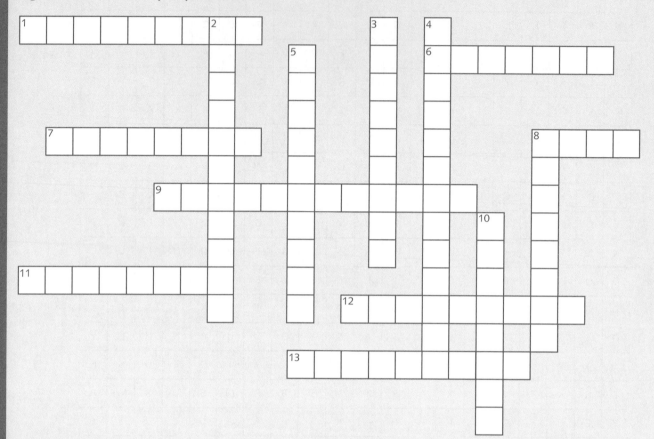

Across

1 A type of cue that originates from a stimulus

6 This type of perceptual experience results in irregularities in perception

7 The term used to describe our perception when we make a perceptual error that causes us to experience an inaccurate perception

8 The visual constancy that is misapplied when viewing the Ames room

9 A perceptual anomaly that results in information taken in by one sense triggering a sensation normally associated with a different sense

11 The type of perceptual experience that results from a limited ability to make sense of information taken in by the senses

12 A stimulus that is able to be interpreted in more than one way

13 A type of cue that does not originate from a stimulus but is associated with the stimulus

Down

2 Synaesthesia is described as this because the person cannot control their synaesthesia

3 A term that describes the way a person experiences the touch of food or beverage in their mouth

4 A perceptual error resulting in the perception of an object or event that has no external reality

5 A type of figure or stimulus that cannot exist in reality

8 Any object or event that evokes a response in someone

10 A consistent perceptual error in interpreting a real external stimulus

11.1 Designing and conducting an experiment

Key knowledge

Investigation design

- psychological science concepts specific to the selected scientific investigation and their significance, including the definition of key terms
- scientific methodology relevant to the selected scientific investigation, selected from classification and identification; controlled experiment; correlational study; fieldwork; modelling; or simulation
- techniques of primary qualitative and quantitative data generation relevant to the investigation
- accuracy, precision, repeatability, reproducibility and validity of measurements in relation to the investigation

Scientific evidence

- the distinction between an aim, a hypothesis, a model and a theory
- ways of organising, analysing and evaluating generated primary data to identify patterns and relationships, including sources of error and remaining uncertainty
- the limitations of investigation methodologies and methods, and of data generation and/or analysis

11.1.1 Writing a testable research question

Key science skills

Develop aims and questions, formulate hypotheses and make predictions

- identify, research and construct aims and questions for investigation

Plan and conduct investigations

- determine appropriate investigation methodology: case study; classification and identification; controlled experiment (within subjects, between subjects, mixed design); correlational study; fieldwork; literature review; modelling; product, process or system development; simulation

Develop

Convert each of the scenarios/topics listed into a research question. Once you have done that, decide which methodology is the most appropriate. An example has been provided.

Example: Restaurants spend a lot of money on lighting design.
Research question: Does light level increase the amount of food consumed?
Methodology: Controlled experiment.

1 Mood and helpful behaviour.

   a Research question:

   _____

   b Methodology:

   _____

2 Praise and student performance.

   a Research question:

   _____

   b Methodology:

   _____

3 Type of cup used to improve the flavour of tea.

   a Research question:

   _____

   b Methodology:

   _____

4 Seating location in a classroom and test results.

   a Research question:

   _____

   b Methodology:

   _____

5 Number of friends and self-esteem.

   a Research question:

   _____

   b Methodology:

   _____

6 Type of feedback given in a sports class.

   a Research question:

   _____

   b Methodology:

   _____

7 Method of data collection (survey or interview) on self-reported attitudes to the change to Aboriginal place names.

   a Research question:

   _____

   b Methodology:

   _____

8 Packaging type on measures of flavour of liquorice.

   a Research question:

   _____

   b Methodology:

   _____

9 Packaging type on measures of cost of lip balm.

   a Research question:

   _____

   b Methodology:

   _____

10 Use of social media on feelings of aloneness.

   a Research question:

   _____

   b Methodology:

   _____

11 Location of mobile phone when studying.

   a Research question:

   _____

   b Methodology:

   _____

12 Using popular influencers to sell products.

   a Research question:

_____

   b Methodology:

_____

13 People with power influence creativity in groups.

   a Research question:

_____

   b Methodology:

_____

14 People exaggerate events depending on who they are retelling it to.

   a Research question:

_____

   b Methodology:

_____

11.1.2 Writing a research hypothesis

Key science skills

Develop aims and questions, formulate hypotheses and make predictions
- identify independent, dependent, and controlled variables in controlled experiments
- formulate hypotheses to focus investigation

Develop

For five research questions you developed in Activity 11.1.1, write an appropriate research hypothesis. In addition, determine the variables under investigation. For a controlled experiment these will be the independent and dependent variables. An example has been provided.

Research question	Does light level affect the amount of food consumed?
If light levels influence food consumption, then people will consume more food in venues with low light levels compared to venues with high light levels.	
Variables	
Independent variable = light levels Dependent variable = amount of food consumed	

1. Research question	
Variables	

2. Research question	
Variables	

3. Research question	
Variables	

4. Research question	
Variables	

5. Research question	
Variables	

11.1.3 What design?

For three research questions in Activity 11.1.1 to which you assigned a controlled experiment as the best type of methodology, decide which research design you will use and why. An example has been provided.

Research question	*Does light level increase the amount of food consumed?*
Methodology	*Controlled experiment*
Research design	*Within group design. This will reduce participant differences and order effects can be controlled by using counterbalancing.*

1. Research question	
Methodology	
Research design	

2. Research question	
Methodology	
Research design	

3. Research question	
Methodology	
Research design	

11.1.4 Sampling procedure

PART A

Underline and identify the population and the sample in the following scenarios.

1   A group of preschool children selected for research into social development; Australian children aged between two and four years.

2   People aged 45 years and over; a group of randomly selected people aged 45 years and over responding to a questionnaire on exercise.

3   Victorian voters registered on the electoral roll; voters responding to a telephone survey on their preferred political leader.

4   Twelve adolescents who are interviewed in a study investigating wellbeing; adolescents in a particular school.

PART B

Matisse, Jonathan and Hayley are three students who are studying VCE. They decide to carry out a research project on attitudes on climate change at their school. They want to find out if students at the school regard climate change as an urgent issue or not. There are a total of 680 students in the school, but they decide to give out surveys to just 40 students to save time. But how to choose which 40 students to survey?

- Jonathan is on the football and basketball team, so suggests handing out the surveys to 40 of his teammates.

- Hayley thinks they should hand out the surveys to students in the library after school.

- Matisse suggests obtaining a list of student emails, and randomly selecting students to receive the survey by email, until 40 surveys are returned.

What are the strengths and limitations of each method?

Table 11.1 Comparison of the strengths and limitations of each method

Method	Strengths	Limitations
Jonathan's		
Hayley's		
Matisse's		

PART C

For three research questions you created in Activity 11.1.1, determine a population of research interest and a method for obtaining a sample. An example has been provided.

Research question	*Does light level increase the amount of food consumed?*
Population	*People over the age of 18 years from one suburb in Melbourne*
Sample	*Invitations sent to 100 random households in a suburb of Melbourne via Australia Post*

1. Research question	
Population	
Sample	

2. Research question	
Population	
Sample	

3. Research question	
Population	
Sample	

11.1.5 Measuring variables

Key science skills

Plan and conduct investigations

- design and conduct investigations; select and use methods appropriate to the investigation, including consideration of sampling technique (random and stratified) and size to achieve representativeness, and consideration of equipment and procedures, taking into account potential sources of error and uncertainty; determine the type and amount of qualitative and/or quantitative data to be generated or collated

Develop

1 For each variable listed in Table 11.2, provide an example of different instruments or techniques that can be used to measure it. Share with the class so you can see that there is more than one way to measure a variable.

Table 11.2 Measuring variables

Variable	Instrument or technique
Self-esteem	
Mood	
Packaging	
Flavour	
Mobile phone location	
Social media	
Sports performance	
Attitude to climate change	
Alcohol consumption	
Reaction time	
Stress	

11.1.6 Presenting summarised data

Adele was very interested in the effects of hugs on physical growth, especially in children. She decided to use rats to model the effects. She designed an experiment that measured the effect of rats being picked up and held for 5 minutes twice a day compared to rats not being picked up at all. One measure she used to determine the effect of hugs was mass, which she measured weekly. In the experiment she used six rats that were split into two groups: hugs (Group A) and no hugs (Group B). She gave them all the same amount of 'food' each day.

Table 11.3 Mass of rats in different conditions

		Rats			
		Week 1	Week 2	Week 3	Week 4
Group A (held)	Rat 1	7.0	8.0	8.5	9.2
	Rat 2	6.0	8.0	9.0	9.7
	Rat 3	7.0	7.0	8.0	9.0
Group B (not held)	Rat 1	7.0	8.0	9.6	11.0
	Rat 2	7.0	8.0	9.9	10.8
	Rat 3	6.0	7.8	10.0	10.6

1 Draw up a summary table that Adele will add into her report using APA conventions.

2 Graph the data showing changes in weight over the four-week period.

3 Write a description of the data.

_____

_____

_____

_____

11.1.7 Understanding errors

Key science skills

Analyse and evaluate data and investigation methods
- identify and analyse experimental data qualitatively, applying where appropriate concepts of: accuracy, precision, repeatability, reproducibility and validity; errors; and certainty in data, including effects of sample size on the quality of data obtained

Develop

1 Use the terms in Table 11.4 to fill in the flowchart in Figure 11.1.

Table 11.4 Terms used to evaluate data

systematic errors	validity
accuracy	precision
reliability	

Figure 11.1 Effect of random and systematic errors on data

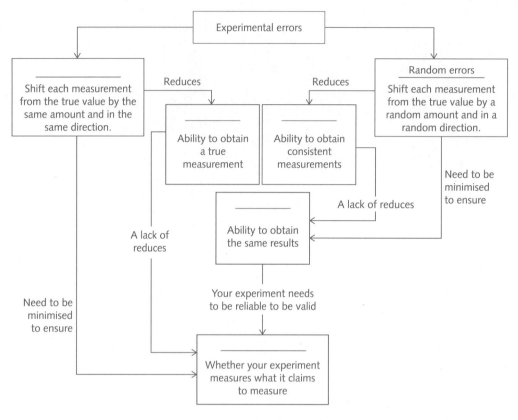

11.2 Science communication

Key knowledge

Science communication
- the conventions of scientific report writing, including scientific terminology and representations, standard abbreviations and units of measurement
- ways of presenting key findings and implications of the selected scientific investigation

11.2.1 Constructing a report

Key science skills

Analyse and evaluate data and investigation methods
- evaluate investigation methods and possible sources of error or uncertainty, and suggest improvements to increase validity and to reduce uncertainty

Construct evidence-based arguments and draw conclusions
- evaluate data to determine the degree to which the evidence supports the aim of the investigation, and make recommendations, as appropriate, for modifying or extending the investigation
- evaluate data to determine the degree to which the evidence supports or refutes the initial prediction or hypothesis
- use reasoning to construct scientific arguments, and to draw and justify conclusions consistent with evidence base and relevant to the question under investigation
- identify, describe and explain the limitations of conclusions, including identification of further evidence required
- discuss the implications of research findings and proposals, including appropriateness and application of data to different cultural groups and cultural biases in data and conclusions

Develop

MATERIALS

- glue
- scissors

PART A

Table 11.5 is a simple introduction for a report on the effects of caffeine on concentration.

Cut out each section. Order each section appropriately to construct a cohesive introduction and paste it onto page 278.

Table 11.5 A simple introduction

Following low (~40 mg) to moderate (~300 mg) caffeine doses, alertness, vigilance, reaction time and attention appear to improve, but less consistent effects are observed on memory and higher-order executive function, such as judgement and decision making (McLennan, 2016).
It is hypothesised that if caffeine influences concentration, then female Year 11 VCE students who consume caffeine will demonstrate better concentration compared to those who do not consume caffeine.
Concentration has been defined as a state in which cognitive resources are focused on certain aspects of the environment rather than on others and the central nervous system is in a state of readiness to respond to stimuli (APA, 2014).
The aim of this study was to determine the effects of caffeine on concentration for female Year 11 VCE students.
Caffeine is a stimulant drug that acts on the brain and nervous system. A wide variety of benefits and risks have been attributed to caffeine, but it is generally agreed that for healthy adults, daily consumption of up to 400 mg (75 kg individual) of caffeine does not present a health risk (McLennan, 2016).
Researchers have equated concentration with sustained attention, and this can be quantified with a sustained attention to a response task, for example, the digit symbol substitution test (Jaeger, 2018).
Caffeine is one of the most widely consumed foods and supplements in the world. Most caffeine is consumed as coffee but caffeine, a psychoactive chemical, is also present in numerous foods and beverages including energy drinks.

PART B

1 Cut out from Table 11.6 the sections of the rest of the report on the effects of caffeine on concentration.

2 Reorder the sections and paste them onto page 279. You will notice that a discussion has not been written. You will need to write a discussion on page 277.

Table 11.6 Elements of a report

Results	*Procedure*:
Acknowledgements	As seen in Table 1, the average time taken to complete the tests when caffeine had not been ingested is 27.2 seconds, compared to the average time taken when caffeine had been ingested which is 31.0 seconds. Figure 1 shows the participants' concentration over time. The participants who didn't have caffeine showed a steadier trend over the 10 trials than the participants who had caffeine.
Method	Both the control group and the experimental group completed the task at the same time of day, in the afternoon.
Participants: Twelve female Year 11 VCE students from one regional school in Victoria who were known to the researcher. A within-groups design was used.	**Table 1:** Average time taken to complete concentration tests, with and without caffeine.
	<table><tr><td>Average time with caffeine (seconds)</td><td>Average time without caffeine (seconds)</td></tr><tr><td>31.0</td><td>27.2</td></tr></table>
Materials (per participant): • 2 × styrofoam cups • caffeinated coffee • decaffeinated coffee • 2 × 250 ml hot water • 2 × digit symbol substitution tests (Jaeger, 2018), see Appendix 1 • stopwatch	The average time taken to complete all 10 trials was calculated. The average time of each trial was graphed to show a trend over time.
The participants were asked to drink a cup of coffee with no sugar within a 10-minute time frame.	

》

Safety and ethical considerations: During the planning of the methodology for this experiment, ethical issues were identified and managed. A copy of the informed consent form can be found in Appendix 2. A risk assessment was conducted using RiskAssess (2022).	APA (American Psychological Association). (2014). Attention. In *APA Dictionary of Psychology*. Apa.org. https://dictionary.apa.org/attention Jaeger, J. (2018). Digit symbol substitution test. *Journal of Clinical Psychopharmacology*, *38* (5), 513–519. https://doi.org/10.1097/jcp.0000000000000941 McLellan, T. M., Caldwell, J. A., & Lieberman, H. R. (2016). A review of caffeine's effects on cognitive, physical and occupational performance. *Neuroscience & Biobehavioral Reviews*, *71* (1), 294–312. https://doi.org/10.1016/j.neubiorev.2016.09.001 RiskAssess. (2022). *RiskAssess – Risk assessments for Australian schools*. Retrieved August 23, 2022 from https://www.riskassess.com.au
Figure 1: Concentration changes over the 10 trials of the digit symbol substitution test for each condition	All participants were then asked to start the digit substitution test (Test 1 in Appendix 1) at the same time. There were 10 trials of the test, and each trial was timed by an assistant researcher using a stopwatch to the nearest tenth of a second.
Discussion	We thank Victoria Jones for the preparation of materials.
Appendix 1: digit symbol substitution test 1 and 2. Appendix 2: informed consent form	**References**
The experiment was then repeated two days later, using decaffeinated coffee, by the same participants and a modified digit symbol substitution test (Test 2 in Appendix 1).	**Appendices**

PART C

1   Write a discussion for the report on the effects of caffeine on concentration.